Principles of Electric Energy Storage Systems for E-Mobility

IEEE Press

445 Hoes Lane

Piscataway, NJ 08854

Principles of Electric Energy Storage Systems for E-Mobility

Sheldon Williamson

Department of Electrical, Computer and Software Engineering
Ontario Tech University
Oshawa
ON
Canada

IEEE Press
WILEY

Published by John Wiley & Sons, Inc., Hoboken, New Jersey.

Library of Congress Cataloging-in-Publication Data has been applied for:

Hardback ISBN: 9781394334094
ePDF ISBN: 9781394334117
ePub ISBN: 9781394334100
oBook ISBN: 9781394334124

Cover Design: Wiley
Cover Image: Pixel Embargo/stock.adobe.com

Contents

About the Authors

Sheldon Williamson (Fellow, IEEE) received the B.E. degree (Hons.) in electrical engineering from the University of Mumbai, Mumbai, India, in 1999, and the M.S. and Ph.D. degrees (Hons.) in electrical engineering from the Illinois Institute of Technology, Chicago, IL, USA, in 2002 and 2006, respectively. He is currently a Professor with the Department of Electrical, Computer, and Software Engineering and the Director of Smart Transportation Electrification and Energy Research Group, Faculty of Engineering and Applied Sciences, Ontario Tech University, Oshawa, ON, Canada. He is a former NSERC Canada Research Chair (Alumni) in Electric Energy Storage Systems for Transportation Electrification. His research focuses on power electronics, battery management systems, electric energy storage, wireless and fast charging, and motor drives for transportation electrification. He has mentored a large and diverse cohort of highly qualified personnel (HQP), including graduate students and postdoctoral fellows now working across academia and industry worldwide. He is an IEEE Fellow and a recipient of the IEEE Bimal Bose Award, the Nagamori Award, and the IEEE PELS Vehicle and Transportation Systems Achievement Award.

V.S.R. Varaprasad Oruganti (Member, IEEE) earned his Ph.D. in electrical engineering from the National Institute of Technology Warangal, India, in 2019. He had fourteen years of teaching and research experience. He currently serves as a Postdoctoral Fellow at the Smart Transportation Electrification and Energy Research Group within the Department of Electrical, Computer, and Software Engineering, Faculty of Engineering and Applied Science, Ontario Tech University, in Oshawa, Ontario, Canada. He received the BHASKARA Advanced Solar Energy Fellowship (2015), awarded by the Indo-US Science and Technology Forum and the Department of Science and Technology, Government of India and POSOCO Power System Award (2017) for his Ph.D. work, jointly presented by Power System Operation Corporation Limited (POSOCO) and the Foundation for Innovation & Technology Transfer (FITT) and the Postdoctoral Fellow Excellence Award (2025) from Ontario Tech University. His current research focuses on advanced power electronic converters for EV charging applications, cybersecurity, and power quality. He is a life member of the Indian Society for Technical Education.

Akash Samanta received the B.Tech. degree in electrical engineering from the West Bengal University of Technology, India, in 2012, the MBA degree in Energy Management from the University of Calcutta, India, in 2014, and the M.Tech. degree in electrical engineering from the University of Calcutta, India, in 2018. From 2014 to 2018, he was a Project Officer and Solar Energy Master Trainer with the Department of Energy Management, Indian Institute of Social Welfare and Business Management, Kolkata, India. In 2025, he received the Ph.D. degree in electrical and computer engineering from Ontario Tech University, Oshawa, ON, Canada. His research interests include Battery Intelligence and Smart Energy Systems.

Latha Anekal received her B.E. in Electronics and Communication Engineering from Visvesvaraya Technological University, Mysore, India, in 2011, and the M.A.Sc. degree in Electrical Engineering from Ontario Tech University, Oshawa, ON, Canada, in 2024. She brings over nine years of industry experience in the electric vehicle and energy storage domain. She is currently pursuing the Ph.D. degree with the Department of Electrical Engineering and is a member of the Smart Transportation, Electrification, and Energy Research (STEER) Group at Ontario Tech University. Her research focuses on advanced battery management systems, electrochemical impedance spectroscopy-based diagnostics, electric energy storage systems, power electronics, and the application of machine learning and artificial intelligence for next-generation electrified transportation systems.

Adhip Sreekumar Kizhuthodi received his BTech in Electrical and Electronics Engineering from TKM College of Engineering, India, in 2013 and his MSc (Engg.) in Electrical Engineering from the Indian Institute of Science in 2017. He has worked as a power system engineer for ABB's and Siemens's electricity market management solutions between 2017 and 2022. He is currently pursuing his PhD at Ontario Tech University.

Chandan Chetri received his B.E. degree (1st class) in Electrical Engineering from Jorhat Engineering College, Assam, India, and his MASc in Electrical Engineering from Ontario Tech University, Oshawa, ON, Canada, in 2020 and 2024, respectively. He is currently pursuing his Ph.D. with the Department of Electrical Engineering and the Smart Transportation Electrification and Energy Research (STEER) Group, Faculty of Engineering and Applied Sciences, Ontario Tech University, Oshawa, ON, Canada. His research interests include electric energy storage systems, power electronic converters, battery management systems, charger equalizers, and battery characterization and estimation using artificial intelligence and machine learning.

Mohit Sharma received his B.Tech. in Electronics and Communications Engineering from the West Bengal University of Technology, Kolkata, India, in 2021. He has over 2 years of industrial experience in automation and control industries. He is currently pursuing his Master of Applied Science with the Department of Electrical Engineering and the Smart Transportation Electrification and Energy Research (STEER) Group, Faculty of Engineering and Applied Sciences, Ontario Tech University, Oshawa, ON, Canada. His current research interests include electric energy storage systems, battery management systems, power electronics converters, the application of machine learning and artificial intelligence in the related field, and drives for different kinds of motors.

Alvin Huynh received the B.E. degree (Hons.) as well as his MASc in electrical engineering from Ontario Tech University (OTU), Oshawa, Canada, in 2021 and 2023, respectively. He is a PhD research student in electrical engineering under the Department of Electrical, Computer, and Software Engineering at OTU. He is curren.

Preface

The rapid evolution of electric mobility is reshaping not only the transportation sector but also the way energy is produced, stored, and utilized. At the center of this transformation lies electric energy storage, an enabling technology that determines the performance, reliability, and sustainability of modern electric vehicles and their interaction with the power grid. In recent years, global electric vehicle adoption has grown at an unprecedented pace, with annual sales surpassing 10 million units and battery demand reaching several hundred gigawatt-hours. This growth underscores the critical importance of efficient, safe, and scalable energy storage solutions. This book has been developed to provide a comprehensive and accessible understanding of this dynamic field.

The content is structured to guide readers from foundational concepts to advanced applications. It begins with an overview of major energy storage technologies, including lithium-ion batteries, supercapacitors, and emerging chemistries that promise higher energy density and improved safety. Subsequent chapters explore the fundamental principles of batteries, covering their internal operation, performance characteristics, and key indicators such as state of charge and state of health. Practical aspects such as battery modeling, thermal management, and pack design are discussed with detailed examples to bridge theory and real-world application.

The book also examines modern charging systems, where fast-charging technologies can reduce charging times to under 30 minutes under appropriate conditions, while raising important considerations related to thermal effects and battery lifespan. In addition, the growing field of second-life battery applications is presented, highlighting how retired EV batteries, which often retain 70–80% of their original capacity, can be repurposed for stationary energy storage, contributing to resource efficiency and sustainability.

In later chapters, attention shifts toward emerging topics such as decentralized energy storage, renewable energy integration, and microgrid applications.

The evolving role of electric vehicles is explored through concepts such as vehicle-to-grid and vehicle-to-home, where aggregated EV batteries can collectively support grid stability and energy balancing. These developments are increasingly important as power systems incorporate higher shares of variable renewable energy.

The final chapter reflects on the challenges and opportunities ahead, including advancements in battery materials, cost reduction, lifecycle management, and intelligent control systems. As the demand for cleaner and more resilient energy systems continues to grow, energy storage will remain at the heart of this transition.

This book is intended for students, researchers, and practicing engineers seeking both theoretical insight and practical understanding of energy storage systems in e-mobility. It is our hope that the material presented here not only informs but also inspires further innovation in this rapidly advancing field.

1

Introduction to Electric Energy Storage Systems

1.1 Introduction

Electric vehicles (EVs) are now recognized as the key technology for decarbonizing road transport, which generates over 15–16% of global CO_2 emissions [1]. EV adoption is surging; global sales reached 14 million units in 2023 and are projected to increase to 70 million units by 2035 [2]. These trends mean that battery packs, the primary energy storage in EVs, are playing a significant role in both transportation and power systems. EV batteries must meet demanding requirements such as extremely high energy density for long driving range, long cycle life, the ability to charge quickly, and rigorous safety and thermal controls [2, 3]. To meet these needs, researchers and manufacturers have rapidly improved lithium (Li)-ion cell chemistries, for example, reducing cobalt (Co) and using nickel (Ni)-rich cathodes and new anode materials, and have driven pack costs steadily downward. For instance, current Li-ion cells now achieve energy densities up to 200–220 Wh/kg and endure thousands of cycles. At the same time, alternative chemistries are being developed, such as sodium (Na)-ion batteries (using abundant Na), which are gaining attention for low-cost, high-cycle storage, and all-solid-state batteries promise yet higher density and improved safety [3, 4].

1.2 Advances in Battery Technologies

Historically, EVs used nickel-metal hydride (NiMH) and even lead-acid batteries, but today Li-ion dominates due to its superior performance. Recent reviews note that modern EV packs typically employ Li nickel-manganese-cobalt oxides (NMC) or Li iron phosphate (LFP) cathodes paired with advanced anodes (often graphite or silicon-based) to maximize energy and power. State-of-the-art automotive Li-ion cells already store about 200–220 Wh/kg and endure well over 2000 cycles.

Research continues to push these limits. For example, silicon-rich anodes and novel solid electrolytes are being investigated to boost capacity and enable 5–10-minute charging. Parallel efforts target cost and sustainability. Na-ion batteries, for instance, leverage inexpensive materials, and emerging recycling processes recover Li, Co, and Ni efficiently. Reviews highlight that next-generation batteries must balance energy density, cost, and lifetime. For example, one recent survey notes that while Ni-based cells once surpassed lead-acid, today's Li-ion chemistries overwhelmingly lead, and future systems like Na-ion, solid-state are being designed to further improve efficiency and resource sustainability [1–4].

EVs also demand high power density for rapid acceleration and fast charging. Advanced battery designs, such as cylindrical, prismatic, or pouch cells, now support extremely high charge/discharge rates, but this in turn places a premium on effective thermal management. Accordingly, much recent work has focused on improving cell stability and heat dissipation. For instance, new electrode materials and electrolytes are optimized for fast-ion transport, and pack designs incorporate advanced cooling (air, liquid, phase-change materials) to prevent overheating during high-rate charging or discharging. These cell- and pack-level innovations, along with economies of scale, have driven the EV battery pack price sharply downward over the last decade, enabling broader adoption [1–4].

1.2.1 Battery Management and Charging Strategies

A battery management system (BMS) is essential for safe and reliable EV operation. The BMS continuously monitors each cell's voltage, current, and temperature and actively controls charging, balancing, and cooling. In practical terms, the BMS prevents thermal runaway and cell imbalance by regulating charge currents, activating cooling loops, and ensuring uniform cell voltages. Reviews emphasize that modern BMS functions include state-of-charge (SOC) and state-of-health (SOH) estimation, often via model-based or data-driven algorithms and real-time fault diagnosis. In addition, the charging strategy is a critical part of system design. Intelligent algorithms adjust the current profile to maximize battery life and accelerate charging when safe. Recent analyses highlight that optimizing the trade-off among fast charging, energy density, and longevity—the so-called "*fast-charging trilemma*"—is a major research theme. In practice, EVs now routinely accept charging rates of more than 150 kW, and ongoing R&D seeks to push this capability to more than 350 kW (providing more than 250 miles of range in approximately 15 minutes) [5].

Large-scale deployment also depends on widespread charging infrastructure and smart charging controls. Home and workplace chargers are the backbone of daily charging, but public and fast-charging networks have expanded rapidly. For example, the IEA reports that public charging points have doubled globally since 2022, exceeding 5 million by 2024. At the same time, digital "*smart charging*" platforms are being rolled out; these systems use grid signals and AI-based controllers to schedule EV charging for low-cost periods or renewable surpluses. According to recent reviews, such smart strategies allow fleets of EVs to be coordinated as a virtual power plant, with vehicles charging when solar and wind energy are abundant and throttling back charging (or even discharging) during system peaks. In short, modern EV charging integration is becoming an active grid resource, not just a passive load [5–7].

1.2.2 Vehicle-to-grid and Second-life Applications

One of the most promising concepts in EV storage is vehicle-to-grid (V2G) operation. In V2G, bidirectional chargers enable an EV to both draw energy from and feed energy back to the grid. Recent reviews highlight that V2G effectively turns every connected EV into a mobile battery, which can absorb excess renewable generation and later discharge during peak demand. This bidirectional energy flow "*spins*" additional virtual inertia into the grid. EV fleets can rapidly inject power to stabilize frequency or function as a spinning reserve during outages. By spreading storage capacity across millions of vehicles, V2G schemes can improve grid flexibility and reduce curtailment of solar/wind energy. Researchers note that enabling technologies such as advanced inverters, communication protocols, and AI-based energy-management systems are key to realizing these benefits [7, 8].

Beyond grid services, EV batteries also have a second life once their vehicular performance degrades below automotive standards, typically when capacity falls to approximately 70–80%. Instead of recycling at once, these used batteries can be repurposed as stationary storage for homes or utilities. In recent times, second-life EV batteries have emerged as a viable solution by repurposing end-of-life batteries, thereby prolonging their lifespan and optimizing resource use. In practice, retired EV packs keep significant capacity and can serve in less-demanding roles, such as backup power, community storage, etc., delaying recycling or disposal. Studies have shown that using second-life batteries in grid storage can cut life-cycle greenhouse gas emissions by approximately 25–56% (compared with producing all-new storage) and reduce system costs by 30–50%. These economic and environmental dividends, along with emerging business models for refurbishing EV packs, are driving interest in second-life markets. Ultimately, both advanced recycling processes and second-life repurposing are seen as crucial for a sustainable battery ecosystem [7, 8]. Through these advances, energy storage is no longer a peripheral concern but a central enabler of electric mobility. In the words of recent reviews, continued innovation in EV batteries in terms of chemistry, management, and integration is "*a prerequisite for a low-carbon energy future*," underpinning both transportation decarbonization and grid reliability.

Global trends from climate change mitigation to transportation electrification further underline why storage innovation is vital. International energy agencies and recent review articles all conclude that continued advances in electrical energy storage (EES) are a prerequisite for a low-carbon energy future. Storage is now recognized not just as an optional upgrade, but as a cornerstone of sustainable transportation and electrification. Through these developments, EES has transformed from a niche auxiliary into a central enabling technology. Current literature repeatedly notes that advancing storage technology in terms of capacity, efficiency, cost, and integration is one of the key requirements for a successful clean energy transition [9].

In summary, the confluence of high-performance battery cells, sophisticated BMS/control strategies, and new paradigms such as V2G and battery reuse is rapidly transforming EVs into versatile energy assets for a clean, flexible power system. In addition to the battery storage, other energy storage technologies also play a significant role in e-mobility applications. Therefore, the classification of various energy storage technologies and their detailed explanations are discussed in the following sections.

1.3 Classification of Energy Storage Technologies

EESs can be systematically classified according to the fundamental form in which energy is stored. As depicted conceptually in Figure 1.1, six principal categories are generally recognized: chemical, electrical, electrochemical, electromagnetic, mechanical, and thermal storage [10–12]. This classification reflects the underlying physical mechanisms governing energy conversion, retention, and release, and it provides a useful framework for comparing technologies, particularly in the context of e-mobility, where performance, weight, and responsiveness are critical.

Each category encompasses a range of technologies with distinct operating principles and practical implications. Chemical energy storage relies on storing energy in molecular bonds, typically in fuels or synthetic carriers. In the context of e-mobility, hydrogen-based systems and synthetic fuels have received renewed attention, especially for heavy-duty transport and long-haul applications. Recent studies highlight that hydrogen fuel cell EVs (FCEVs) offer rapid refueling and long driving ranges, although challenges related to energy-storage density, infrastructure, and system cost remain significant.

EES, by contrast, involves the direct storage of charge in electric fields. Technologies such as capacitors and supercapacitors fall into this category. While their energy density is relatively low compared with that of batteries, their extremely high-power density and fast response times make them attractive for specific vehicular functions. Recent literature emphasizes their growing role in

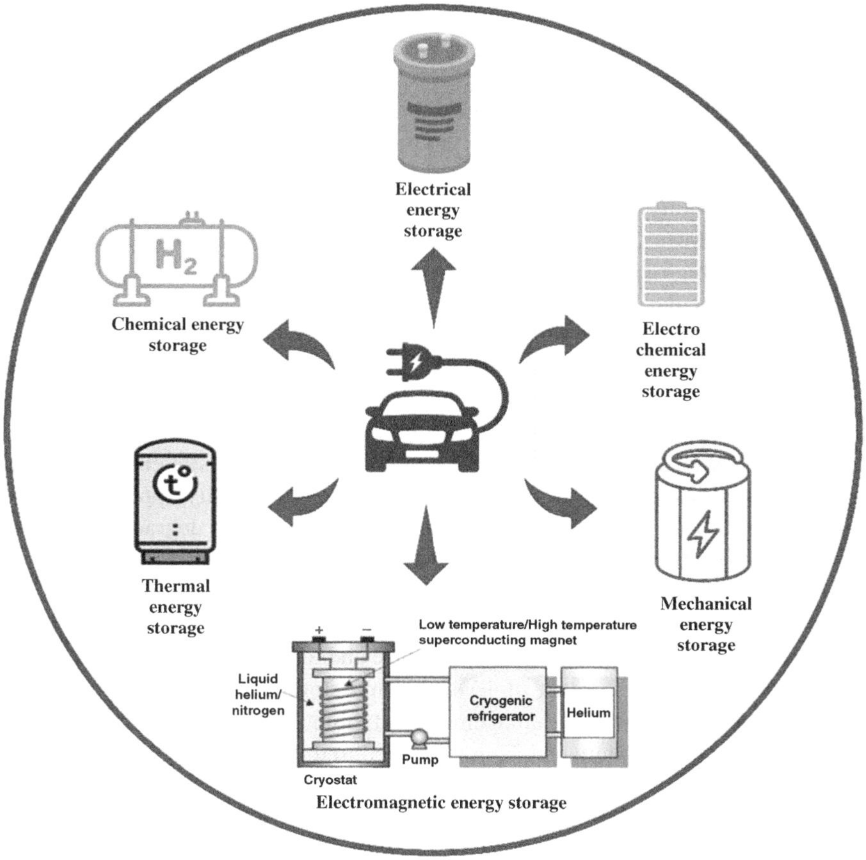

Figure 1.1 Energy storage technologies.

hybrid energy storage systems (HESS), where supercapacitors are paired with Li-ion batteries to handle peak power demands, such as acceleration and regenerative braking, thereby reducing battery degradation and improving overall system lifespan [1–12].

Electrochemical storage represents the cornerstone of modern e-mobility. It is based on reversible redox (reduction–oxidation) reactions, as seen in batteries and flow cells. Li-ion batteries continue to dominate EV applications due to their high-energy density, efficiency, and declining cost. Since 2021, significant advancements have been reported in lithium iron phosphate (LFP), Ni-rich chemistries (nickel manganese cobalt [NMC]/nickel cobalt aluminum [NCA]), and emerging solid-state batteries. These developments aim to enhance safety, extend driving range, and reduce reliance on critical materials. Na-ion batteries have also gained momentum as a promising alternative for cost-sensitive EV segments, with recent prototypes proving competitive performance for short- to high-range mobility applications.

Electromagnetic energy storage uses magnetic fields to store energy, most notably in superconducting magnetic energy storage (SMES) systems. Although SMES offers near-instantaneous response and extremely high efficiency, its application in e-mobility remains limited due to the need for cryogenic cooling and high costs. However, recent research has explored its potential in niche applications such as high-performance transport systems and grid–vehicle interaction scenarios, particularly where ultra-fast power delivery is needed.

Mechanical energy storage involves storing energy in the form of motion or position. Technologies such as flywheels and elastic (spring-based) systems fall under this category. In e-mobility, flywheel systems have been revisited for regenerative braking in buses, trams, and high-performance vehicles, where their high-power density and long cycle life are helpful. Elastic energy storage, although less mature, has been investigated for micro-mobility applications and auxiliary systems, offering rapid energy release with minimal degradation. These systems are particularly relevant for short-duration, high-power events rather than sustained energy supply.

Thermal energy storage stores energy in the form of heat, either as sensible heat (temperature change), latent heat (phase change), or through thermochemical processes. In the context of EVs, thermal storage has gained importance primarily for battery thermal management and cabin climate control. Additionally, waste heat recovery systems are being explored to improve overall vehicle energy efficiency, especially in cold climates where heating demands significantly impact driving range [10–12].

The detailed classification shown in Figure 1.2 further expands on these categories, illustrating the diversity of technologies within each class. Importantly, these storage options differ significantly in terms of key performance metrics such as energy density, power density, efficiency, lifecycle, cost, and technological maturity. For e-mobility applications, no single technology satisfies all requirements; instead, recent research increasingly supports hybridization strategies, combining complementary storage systems to optimize performance, durability, and cost-effectiveness.

1.3.1 Chemical Energy Storage

Chemical energy storage involves storing energy within the chemical bonds of substances and releasing it later through chemical reactions. In this approach, electricity is often surplus power from renewable sources that is first converted into chemical energy using processes such as electrolysis or chemical synthesis.

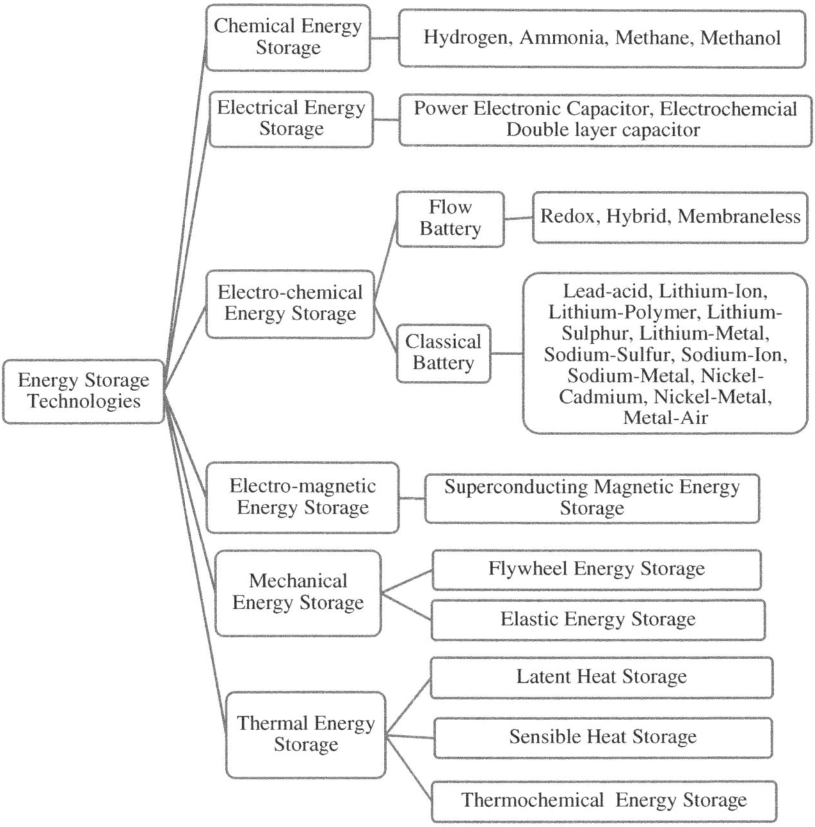

Figure 1.2 Classification of electric energy storage technologies.

The stored energy can then be recovered as electricity or heat through fuel cells, engines, or combustion systems. Common chemical energy carriers discussed in recent literature include hydrogen (H_2), ammonia (NH_3), methane (CH_4), and methanol (CH_3OH). In recent years, energy organizations have highlighted chemical storage as an important solution for long-duration and large-scale energy storage, particularly where batteries are less suitable [13–15].

1.3.1.1 Hydrogen (H_2)

Hydrogen is one of the most widely studied chemical energy carriers. It is mainly produced either by electrolysis of water using electricity or by steam reforming of natural gas. When produced using renewable electricity, hydrogen is considered a low-carbon or "green" fuel. Hydrogen has a very high gravimetric energy density, approximately 33 kWh/kg, based on its lower heating value, which means it stores

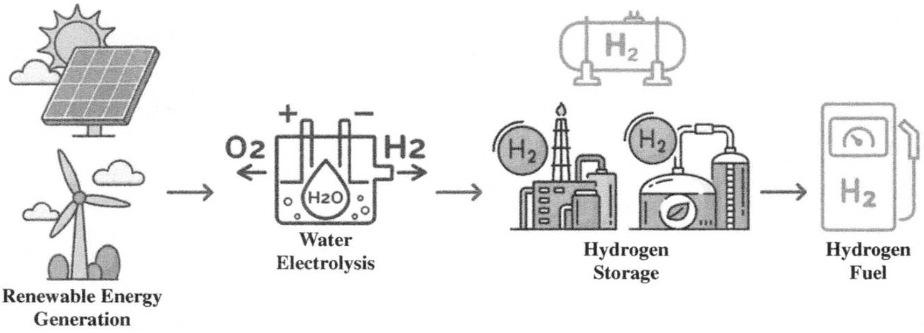

Figure 1.3 Hydrogen storage application.

a large amount of energy per unit mass. However, its volumetric energy density is very low. Even when compressed to high pressures, such as 700 bar, gaseous hydrogen contains only about 5 MJ/L, which is much lower than many liquid fuels. The basic illustration of hydrogen storage is shown in Figure 1.3.

Hydrogen can be stored in several forms, including as a high-pressure gas, a cryogenic liquid, or within solid materials such as metal hydrides. Each of these storage methods presents technical challenges. High-pressure storage requires strong and expensive tanks, liquid hydrogen suffers from boil-off losses due to its very low boiling point, and metal hydrides add weight and cost while facing material degradation issues. Another challenge discussed in recent studies is hydrogen embrittlement, where hydrogen weakens metals used in pipelines and storage vessels. When used for power generation, hydrogen is commonly converted back into electricity using fuel cells, which typically achieve efficiencies of about 50–60%. Hydrogen is considered a strong candidate for long-duration and seasonal energy storage, as well as for long-distance energy transport. However, current research also points out its limitations, including relatively low overall round-trip efficiency and the lack of widespread infrastructure for hydrogen production, storage, and distribution [12, 13].

1.3.1.2 Ammonia (NH_3)

Ammonia is produced by combining nitrogen and hydrogen, usually through the Haber–Bosch process. Because it does not contain carbon, ammonia is considered a carbon-free fuel at the point of use. It has a hydrogen content of about 17.6% by weight and can be stored as a liquid under relatively mild conditions, around 1 MPa at room temperature. This makes ammonia easier to store and transport than hydrogen. Liquid ammonia has a volumetric energy density of about 10.5 MJ/L, which is higher than that of liquid hydrogen (8.49 MJ/L).

Ammonia can be used as an energy carrier in two main ways: it can be cracked back into hydrogen for use in fuel cells, or it can be burned directly in engines or gas turbines. One major advantage of ammonia is its long history of

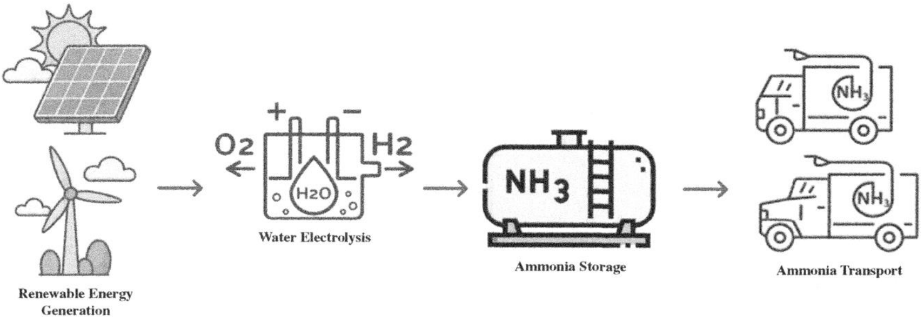

Figure 1.4 Ammonia synthesis and transportation.

large-scale industrial use, particularly in fertilizer production, which has lead to the development of well-established storage and transport technologies. The ammonia synthesis and transportation process flow diagram is illustrated in Figure 1.4. Recent literature also highlights its relatively easy liquefaction compared to hydrogen [14].

However, ammonia has several notable disadvantages. It is toxic and must be managed carefully to avoid health and environmental risks. In addition, energy losses occur during its synthesis and during cracking back into hydrogen, leading to lower overall round-trip efficiency. At present, most ammonia-based applications are still at the pilot or demonstration stage. Economic analyses published in recent years suggest that ammonia could become cost-competitive with conventional fuels if low-cost, low-carbon hydrogen becomes widely available.

1.3.1.3 Methane (CH_4)

Methane stores energy in carbon—hydrogen bonds and is commonly used in the form of natural gas. It can also be produced synthetically through "power-to-gas" processes, in which renewable electricity is used to convert carbon dioxide and hydrogen into methane. The methane production and use process is illustrated in Figure 1.5. Methane has a gravimetric energy density of about 38 MJ/kg. On a volumetric basis, compressed methane has an energy density comparable to or slightly lower than that of ammonia; however, it typically requires higher storage pressures.

One advantage of methane is that it can be transported and distributed using existing natural gas pipeline infrastructure. It can also be used efficiently in combined-cycle gas turbines or in certain types of fuel cells. However, methane combustion or reforming releases carbon dioxide unless combined with carbon capture and storage technologies. Recent studies say that producing synthetic methane at a large scale remains expensive, especially when carbon-neutral pathways are used, which limits its current deployment as a clean energy-storage choice [15, 16].

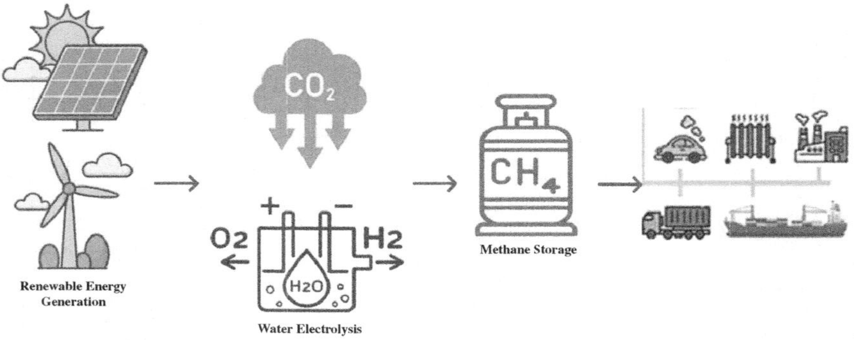

Figure 1.5 Methane production and utilization.

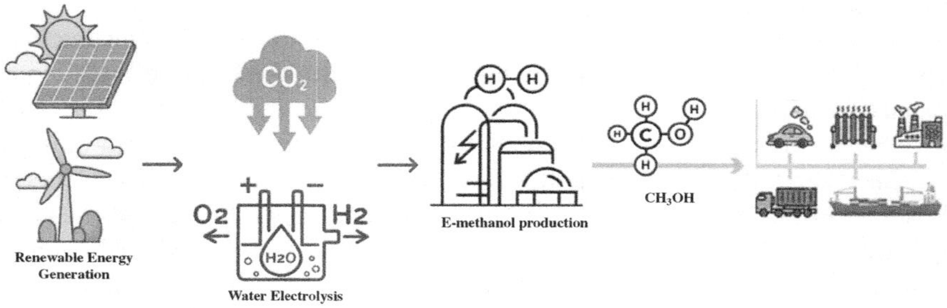

Figure 1.6 Methanol production and utilization.

1.3.1.4 Methanol (CH_3OH)

Methanol is a liquid organic hydrogen carrier and has gained increasing attention in recent research. It has an energy density of about 15.8 MJ/L, or roughly 4.4 kWh/L, which is significantly higher than that of compressed hydrogen. Because methanol is a liquid at ambient conditions, it is easy to store, transport, and handle using existing fuel infrastructure. It can be used directly as a fuel in combustion engines or in direct methanol fuel cells [17]. The process flow of methanol production and utilization is shown in Figure 1.6.

Recent studies highlight that methanol synthesis can consume carbon dioxide, offering a pathway toward carbon-neutral fuels when renewable hydrogen is used. This feature makes methanol attractive for integrating energy storage with carbon recycling. However, methanol is toxic and poses health risks if improperly managed. In addition, methanol fuel cells generally have lower efficiencies compared to hydrogen fuel cells, and some carbon emissions still occur depending on the production and use pathway.

In summary, chemical energy storage systems provide very high energy densities, especially in terms of mass, and are well suited for long-duration and

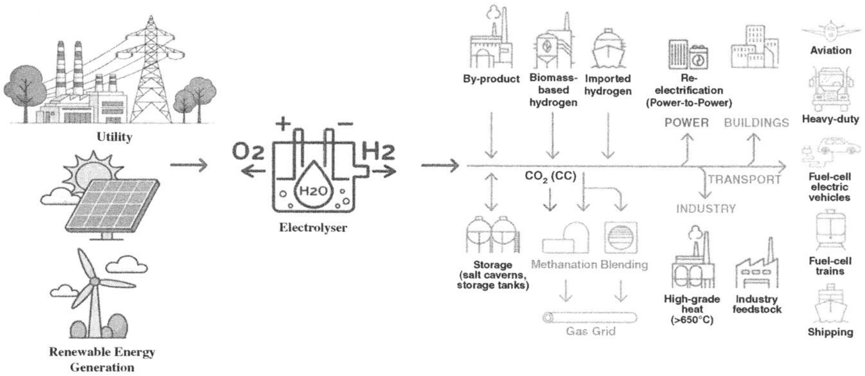

Figure 1.7 Chemical energy storage and utilization. *Source*: Adapted from [13].

seasonal energy storage. These characteristics make them attractive for E-mobility, large-scale grid applications, and energy-intensive industries. However, their overall round-trip efficiencies are typically modest, often ranging from 20% to 50%, depending on the specific conversion processes involved. Capital costs remain high, although ongoing research and pilot projects are steadily improving performance and reducing costs. A detailed overview of chemical energy storage and utilization is shown in Figure 1.7.

Hydrogen and natural gas are already well established, while ammonia and methanol are at earlier stages of commercialization, mainly demonstrated through research projects and pilot-scale applications. Recent literature consistently identifies chemical energy storage as a critical component of future low-carbon energy systems [13–17].

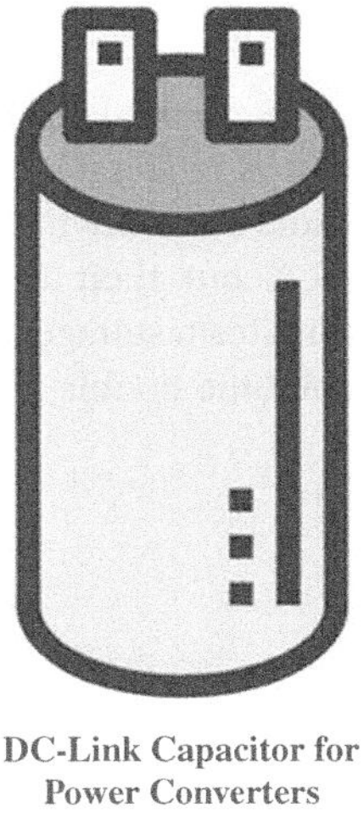

Figure 1.8 DC-link capacitor used in power converters and drives.

1.3.2 Electrical Energy Storage

EES devices store charge in electric or electrostatic fields. There are two main categories:

1. Power electronic capacitors (dielectric capacitors used in power electronics)
2. Electrochemical double-layer capacitors (EDLCs), or supercapacitors.

Power Electronic Capacitors: These are conventional film or electrolytic capacitors like DC-link capacitors used in power converters and drives, as illustrated in Figure 1.8.

They operate at high voltage and can deliver very high-power bursts, short-term buffering of switching transients. However, their energy storage is *negligibly small* (specific energy of about 10^{-5} MJ/kg), so they only smooth out disturbances on micro- to millisecond timescales. In practice, these capacitors are mature, low-cost, and dependable, but they cannot serve as standalone energy storage for more than a brief ride-through or smoothing functions [18, 19].

Electrochemical Double-layer Capacitors (Supercapacitors): These use high-surface-area carbon electrodes in an electrolyte to store charge in the electrical double layer. Supercapacitors have **exceptional power density** and extremely fast charge/discharge (seconds), as well as very long cycle life (>10**6** cycles). The basic configuration of an EDLCs is illustrated in Figure 1.9 [19]. For example, they can buffer regenerative braking pulses or ride through short outages. However, their energy density is much lower than batteries, typically about 1–10 Wh/kg (0.004–0.036 MJ/kg) compared to approximately 100–250 Wh/kg for modern Li-ion batteries.

The round-trip efficiency of supercapacitors is very high (often >98%), and their cost is moderate. In summary, EDLCs are commercially mature for high-power, short-duration buffering (transit regenerative braking, uninterruptible power, etc.), but their low energy capacity means they are not used for sustained long-term storage. Many researchers are developing *hybrid* "ultracapacitors" that combine double-layer and pseudocapacitive effects to boost energy density, but

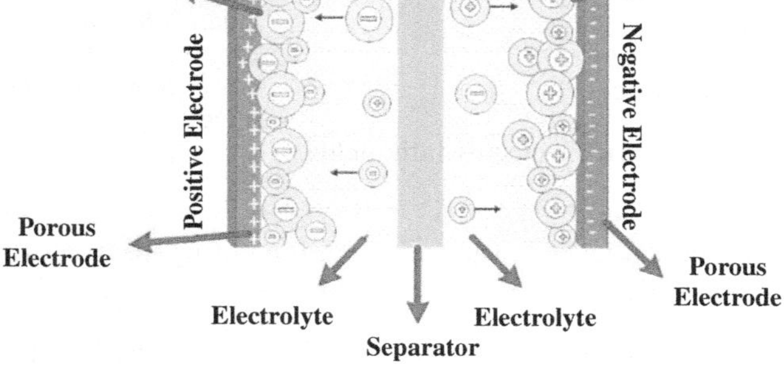

Figure 1.9 Electrochemical double-layer capacitor configuration. *Source:* Adapted from [19].

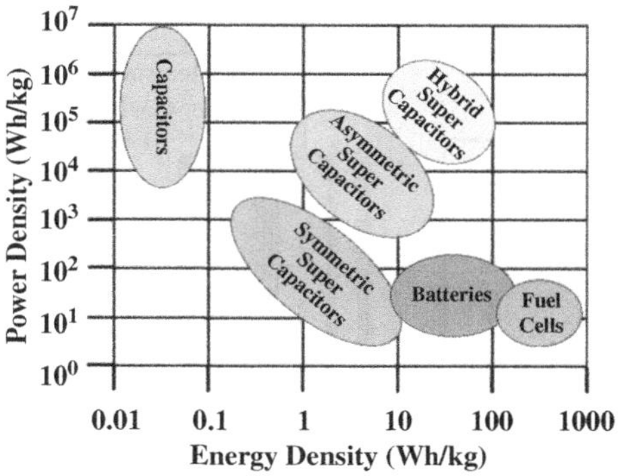

Figure 1.10 A Ragone plot comparing capacitors, supercapacitors, and batteries.

the basic EDLC is still a power-oriented device. A Ragone plot comparing capacitors, supercapacitors, and batteries, to illustrate energy vs. power density [20], is depicted in Figure 1.10.

1.3.3 Electrochemical Energy Storage

Electrochemical energy storage converts electrical energy via reversible chemical reactions. The two main forms are flow batteries and solid-electrode batteries. The latter are conventional rechargeable batteries [21, 22].

1.3.3.1 Flow Batteries

Flow batteries store energy in liquid electrolytes that circulate through electrochemical cells. The energy capacity is set by the volume and concentration of the active species in external tanks, while the power is decided by the size of the cell stack. Major types include:

- **Redox Flow Batteries (RFBs):** A classic example is the vanadium redox flow battery (VRFB), which uses V^{2+}/V^{3+} on one side and V^{4+}/V^{5+} on the other in sulfuric acid [23]. During charge/discharge, the vanadium ions shuttle electrons across the cell as shown in Figure 1.11. RFBs have the advantage of decoupled energy and power (tank size vs. stack area), making them easy to scale. They also have excellent cycle life (>10,000 cycles) and allow deep discharge without damage. Contemporary VRFBs achieve only modest energy densities (typically <25 Wh/kg) due to solubility limits, which corresponds to 20–40 Wh/L for large tank systems. They are inherently safe and nonflammable. The capital cost of VRFB systems is moderate and improves at large scale, though pumps and thermal management add complexity. Commercial VRFB installations (hundreds of kW to MW scale) exist for grid storage, but high vanadium cost and system complexity have

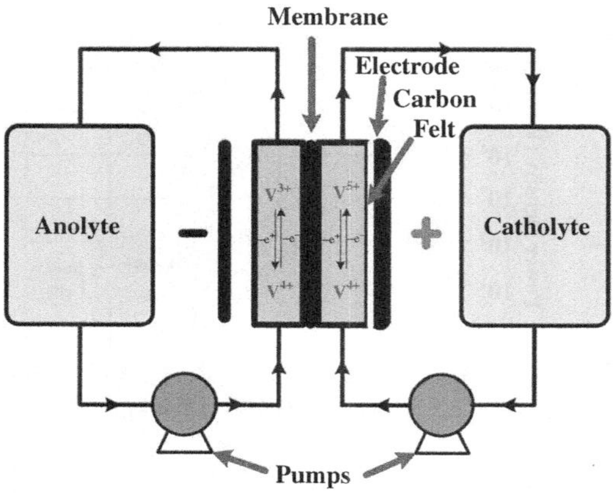

Figure 1.11 Vanadium redox flow battery configuration.

spurred research into alternative chemistries (e.g., iron–chromium, organic redox couples, hybrid designs).

- **Hybrid Flow Batteries:** These combine a circulating electrolyte on one side with a static (often plating) electrode on the other. For example, zinc–bromine or zinc–iron flow batteries use a flowing bromide or iron solution with a zinc metal electrode that plates on charge. Such hybrids can achieve higher energy density (because the plated metal holds more charge), but they introduce challenges, including zinc dendrite formation and require separators. Cutting-edge lab research includes "membraneless" microfluidic flow cells, but these are still experimental [24, 25]. The basic configuration of the zinc–bromine flow battery is shown in Figure 1.12.

Overall, flow batteries are well suited to multi-hour energy storage (tens to hundreds of kWh), providing long service life and flexible scaling. Their main drawback is lower volumetric/gravimetric energy density and higher system cost compared to batteries. VRFBs are the most mature with commercial products available, whereas many alternative flow chemistries are still in demonstration or development stages.

1.3.3.2 Classical Batteries

This category covers "solid-electrode" rechargeable batteries, which use fixed electrodes and internal electrolytes [26, 27]. Key chemistries include:

- **Lead–Acid Batteries:** The oldest rechargeable battery (invented 1859). Lead–acid cells store energy via $Pb/PbSO_4$ at both electrodes. They have very low specific energy (approximately 35–40 Wh/kg) and low energy density

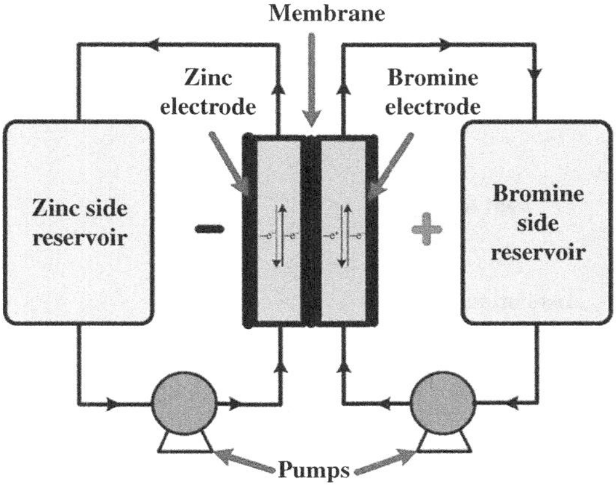

Figure 1.12 Zinc–bromine flow battery configuration.

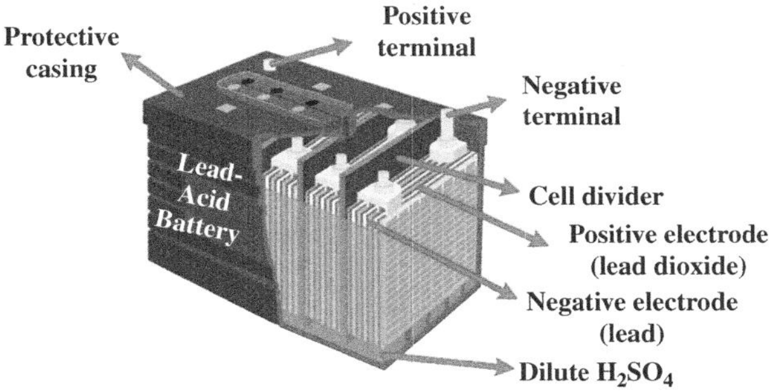

Figure 1.13 Lead–acid battery configuration.

(approximately 80–90 Wh/L), making them heavy for the energy stored. On the plus side, lead–acid batteries are inexpensive, robust, and have mature recycling. They can deliver high surge currents used for engine starters, but only a few hundred deep cycles (≈300–500) and limited calendar life. Efficiency is modest (approximately 75–85%). Because of their low energy density, lead–acid is largely confined to starter, backup, and stationary applications, rather than EV propulsion or portable devices [28]. The basic configuration of lead-acid battery is shown in Figure 1.13.

- **Nickel–Cadmium (Ni–Cd) Batteries**: Ni–Cd batteries use a Ni hydroxide cathode and a cadmium anode with KOH electrolyte. They have moderate energy density (approximately 50–75 Wh/kg) and about 150–300 Wh/L. Ni–Cd cells tolerate abuse (overcharge, deep discharge) well and can sustain high discharge rates; they can last 1500–2500 cycles. However, cadmium is toxic, and Ni–Cd suffers from the “memory effect” and weight issues.

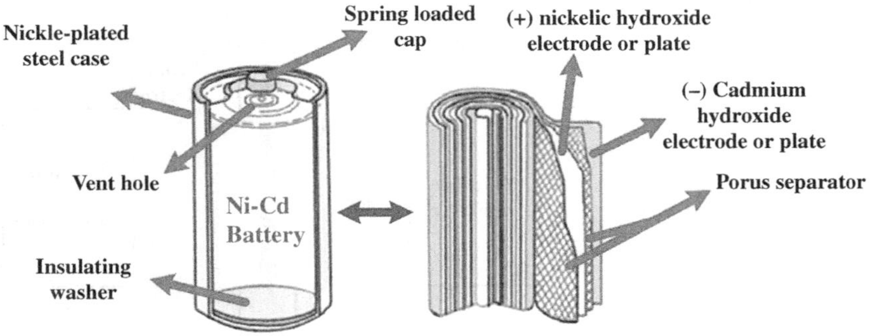

Figure 1.14 Nickel–cadmium battery configuration.

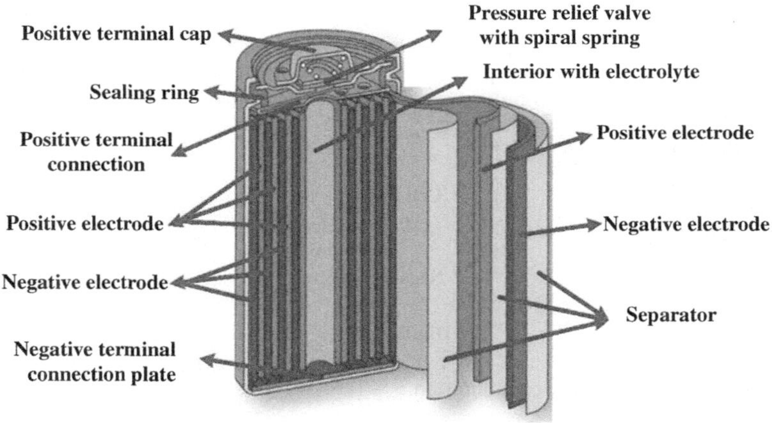

Figure 1.15 Nickel–metal hydride battery configuration.

Consequently, Ni–Cd has largely been replaced by Ni–MH in consumer devices [29]. The Ni–Cd battery configuration is shown in Figure 1.14.

- **Nickel–Metal Hydride (Ni–MH) Batteries**: Ni–MH cells (Ni hydroxide cathode, hydrogen-absorbing alloy anode) have higher energy density than Ni–Cd (approximately 70–100 Wh/kg) and 140–300 Wh/L. They achieve a few hundred to approximately 1000 cycles and have better charge retention. Ni–MH was popular in hybrid vehicles (e.g., Toyota Prius) and cordless tools. Today, Ni–MH has been largely superseded by Li-ion, due to Li-ion's much higher energy and lower long-term cost [30]. The Ni–MH battery configuration is illustrated in Figure 1.15.
- **Li-ion Batteries**: The Li-ion batteries are the dominant rechargeable battery technology today, with many chemistries (e.g., $LiCoO_2$, NMC, $LiFePO_4$) is

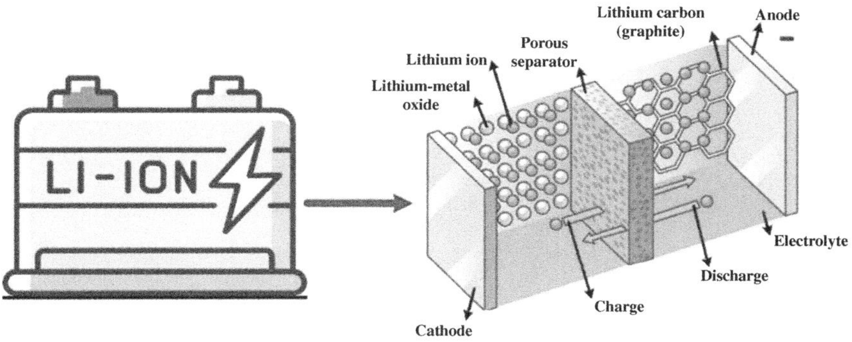

Figure 1.16 Li–ion battery configuration.

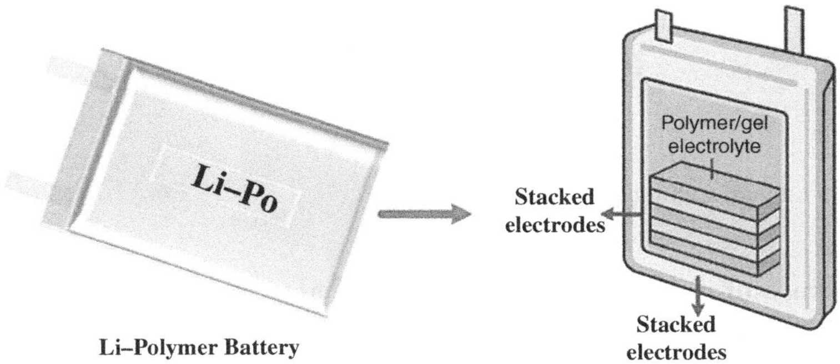

Figure 1.17 Lithium–polymer battery configuration.

Li-ion. The basic configuration of a Li-ion battery is shown in Figure 1.16. Commercial Li-ion cells typically achieve about 100–250 Wh/kg (0.36–0.90 MJ/kg) and 250–700 Wh/L. Coulombic efficiency is very high (approximately 99%), and round-trip efficiencies are approximately 90–95%. Cycle life depends on chemistry and conditions, but is often >1000 cycles; advanced designs now exceed 2000 cycles. Li-ion batteries are relatively expensive and require careful management for safety, as they can undergo thermal runaway. Resource limitations (Li, Co, Ni) are a consideration. Nonetheless, Li-ion batteries are the standard technology for portable electronics, EVs, and utility-scale storage [31].

- **Lithium–Polymer (Li–Po) Batteries**: A variant of Li-ion using a polymer or gel electrolyte. Performance (energy density, power) is like Li-ion, but cells can be made thinner and more flexible [31]. Li–Po is used in drones, wearable devices, and mobile devices where form factor is important. The Li–Po battery has been illustrated in Figure 1.17.

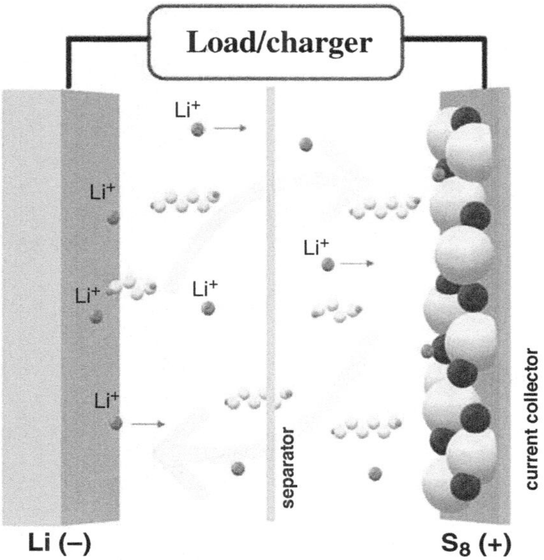

Figure 1.18 Lithium–sulfur battery configuration.

- **Lithium–Sulfur (Li–S) Batteries**: These batteries use a Li metal anode and S cathode as illustrated in Figure 1.18. The theoretical specific energy is extremely high (approximately 2600 Wh/kg of Li + S), so practical prototypes aim approximately 500–600 Wh/kg. Lab cells have achieved approximately 250–350 Wh/kg, and S is cheap and abundant. The main challenge is the "polysulfide shuttle" that causes rapid capacity fade, cycle life is currently poor. Li–S technology is still in the research or pilot stage [32].
- **Lithium–Metal Batteries**: Using Li metal as the anode (paired with various cathodes) can push the energy density even higher (theoretical cell-level energy densities of up to approximately 500–1000 Wh/kg of cell) because Li metal is lighter than graphite or silicon. Recent R&D (often with novel solid or highly concentrated electrolytes) has produced Li-metal prototypes in the approximately 500 Wh/kg range. Improving cycle life is an active area of research; new electrolytes have demonstrated Li-metal cells surviving hundreds to thousands of cycles (e.g., 483 cycles at 500 Wh/kg, and reports of approximately 3000 cycles at 80% capacity). Commercial Li-metal batteries are not yet available, but this is a "holy grail" for next-generation high-energy cells [32]. The basic configuration of a lithium–metal battery is shown in Figure 1.19.

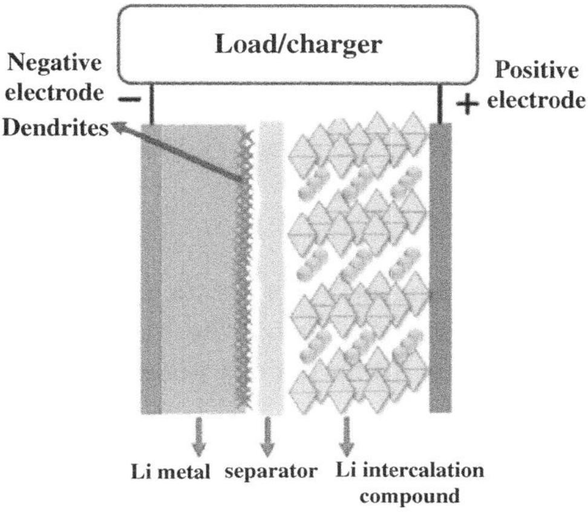

Figure 1.19 Lithium–metal battery configuration.

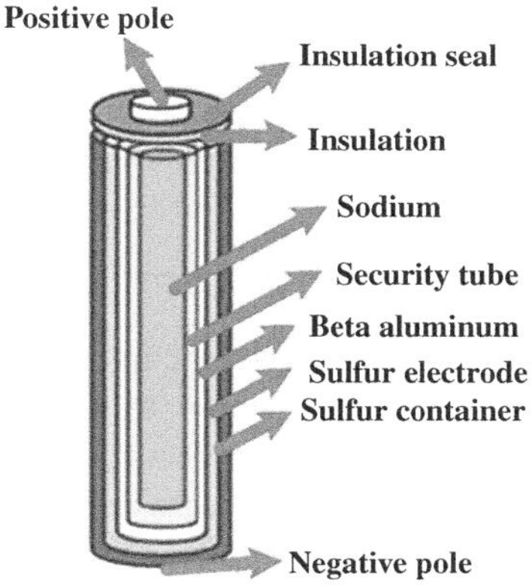

Figure 1.20 Sodium–sulfur battery configuration.

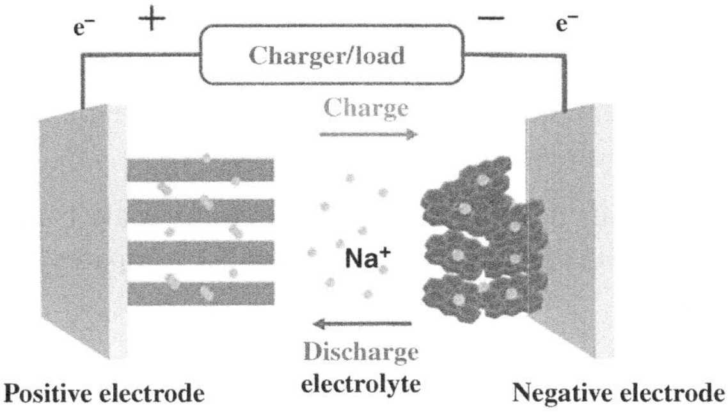

Figure 1.21 Sodium-ion battery configuration.

- **Sodium–Sulfur (Na–S) Batteries:** These are high-temperature (300–350 °C) batteries with molten Na and S electrodes and a beta-alumina electrolyte. The basic configuration of a Na–S battery is shown in Figure 1.20. Na–S batteries can reach approximately 150 Wh/kg and have a very long life (>2500 cycles). They have been deployed in grid-scale energy-storage applications. The drawbacks are the high operating temperature and the need for thermal insulation. Owing to their complexity, Na–S batteries are used in large stationary installations rather than EVs [28].
- **Sodium-ion Batteries (Na-ion):** An emerging "drop-in" battery using Na^+ instead of Li^+. Laboratory Na-ion cells have achieved energy densities of approximately 90–160 Wh/kg (lower than Li-ion but improving). The basic schematic of a Na-ion battery is shown in Figure 1.21.

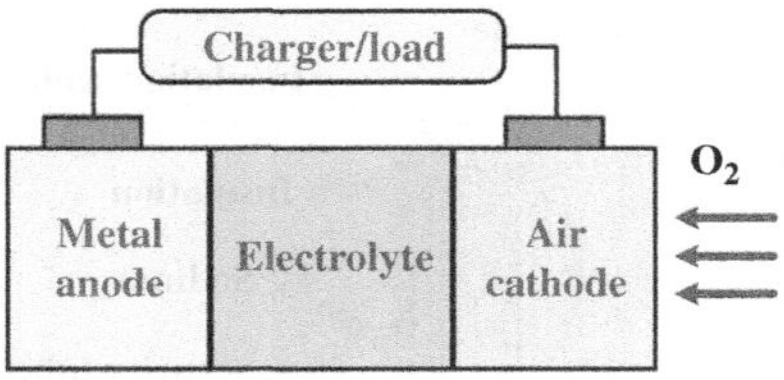

Figure 1.22 Metal–air battery configuration.

Advantages include lower cost and abundant materials such as Na and alumina. Na-ion technology has only recently entered the market, with limited commercialization since 2022, primarily for stationary energy-storage systems and two-wheeler EVs.

- **Metal–Air Batteries (Zn–Air, Li–Air, etc.):** These batteries use a metal anode (Zn, Li, Al, etc.) and oxygen from air as the cathode reactant. The theoretical energy densities are extremely high (e.g., Zn–Air theoretical approximately 1086 Wh/kg; Li–Air up to approximately 11,400 Wh/kg). In practice, most metal–air cells are primary (nonrechargeable); for example, Zn–air hearing-aid cells achieve very high energy per weight by venting oxygen. Rechargeable metal–air systems face significant challenges (air electrode catalysis, electrolyte management, dendrites, CO_2 contamination, etc.) [33]. The basic schematic of a metal–air battery is shown in Figure 1.22.

Some technologies such as zinc–air flow batteries and laboratory-scale Li–O_2 cells, are being studied but none are yet practical at scale. Table 1.1 compares key metrics for major battery types.

Advantages and Limitations: Classical battery technologies generally involve trade-offs among energy density for power and cost. Li-ion batteries offer the highest energy density and efficiency of the mature chemistries. Lead-acid and Ni–Cd batteries are well-proven and inexpensive but are heavy with low energy. Ni–MH batteries provide better safety and cycle life than Ni–Cd batteries, but at a higher cost. Na-based (Na–S, Na-ion) and metal–air batteries promise to use abundant materials or ultra-high theoretical energy, but they face technical hurdles (high temperature, limited cyclability, etc.). Modern Li-ion pack costs are about \$100–200 per kWh (e.g., ≈\$115/kWh in 2024). Battery efficiencies are generally high (>80–90%), but all batteries have finite cycle life (hundreds to a few thousand cycles) and require proper management for safety (especially Li-based chemistries). Research continues on advanced chemistries (solid-state, metal anodes, Na systems) to improve energy, life, and cost.

1.3.4 Electromagnetic Energy Storage

Electromagnetic storage uses magnetic fields or magnetization.

Table 1.1 Comparison of selected energy-storage technologies.

Battery type	Specific energy	Energy density	Power density	Cycles	Cost	Status
Lead–Acid	35–40 Wh/kg	80–90 Wh/L	Moderate	<500	Low	Mature (automotive, Uninterruptible Power Supply)
Ni–Cd	40–60 Wh/kg	50–150 Wh/L	High	~2000	Medium	Mature (tools, standby)
Ni–MH	60–120 Wh/kg	140–300 Wh/L	High	500–1000	Medium	Commercial (hybrids, electronics)
Li-ion	100–250 Wh/kg	250–730 Wh/L	High	>1000	Medium–high	Commercial (EVs, portable)
Li-polymer	Same as Li-ion	—	—	—	—	Commercial (consumer electronics)
Li–S (research)	~200 laboratory-scale 300 Wh/kg	—	Moderate	<100 (lab)	Low (materials)	Emerging (pilot)
Li–metal (solid-state) (R&D)	500–600 Wh/kg	—	High	TBD	Low (Li amt.)	Experimental
Na–S (300 °C)	~150 Wh/kg (64 Wh/L)	Moderate	High	>2500	Low	Commercial (grid)
Na–ion	~100–160 Wh/kg	—	High	TBD	Low	Emerging
Zn–air (primary)	≈300 Wh/kg theoretical	(High, depends)	Moderate	1 (nonrechargeable)	Low	Niche (hearing aids)
Supercapacitor	1–10 Wh/kg	3–10 Wh/L	Very high	>100,000	High	Commercial
SMES	~1–11 Wh/kg	~<40 kJ/L (<11 Wh/L)	Extremely high	>10^6	Very high	Demonstration

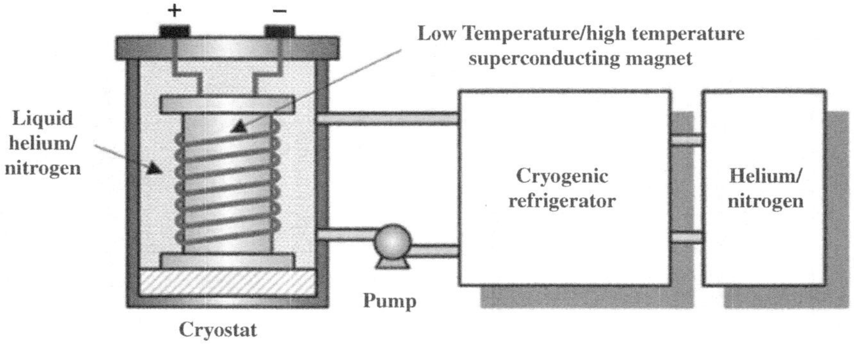

Figure 1.23 Schematic of the SMES system.

1.3.4.1 Superconducting Magnetic Energy Storage

Superconducting magnetic energy-storage (SMES) stores energy in the magnetic field of a superconducting coil. The basic schematic of the SMES system is illustrated in Figure 1.23. Under cryogenic conditions (below T_c), the coil has zero resistance; therefore, power can be exchanged virtually loss-free. SMES systems can release enormous power almost instantaneously (with a power density approximately 10^4–10^5 kW/kg) and achieve a high round-trip efficiency (approximately 95%). However, their energy capacity is very low due to the limited coil inductance and cryostat size. The typical SMES energy density is <40 kJ/L (≈11 Wh/L). Ongoing advances in high-T_c superconductors aim to reduce costs and refrigeration needs [34].

1.3.5 Mechanical Energy Storage

Mechanical energy storage in vehicles includes high-speed flywheels and elastic devices such as springs [35, 36]. The basic flywheel mechanism is illustrated in Figure 1.24.

Modern flywheel systems spin rotors in a near-vacuum using magnetic bearings, yielding very high-power bursts with minimal wear. In EVs, flywheels have been explored for regenerative braking or hybrid powertrains: they can absorb and release energy almost instantaneously and endure millions of cycles. For example, hybrid locomotives and some heavy vehicles use flywheel modules to capture braking energy. However, flywheels store relatively little energy per mass (tens of Wh/kg at best), so they are suited only to short-duration pulses (seconds to minutes). They complement batteries by smoothing power spikes in UPS, train braking or freight applications. Elastic spring-based storage has also been explored in recent times. Elastic energy storage, particularly spring-based systems, is gaining renewed attention in e-mobility as a lightweight, high-power solution

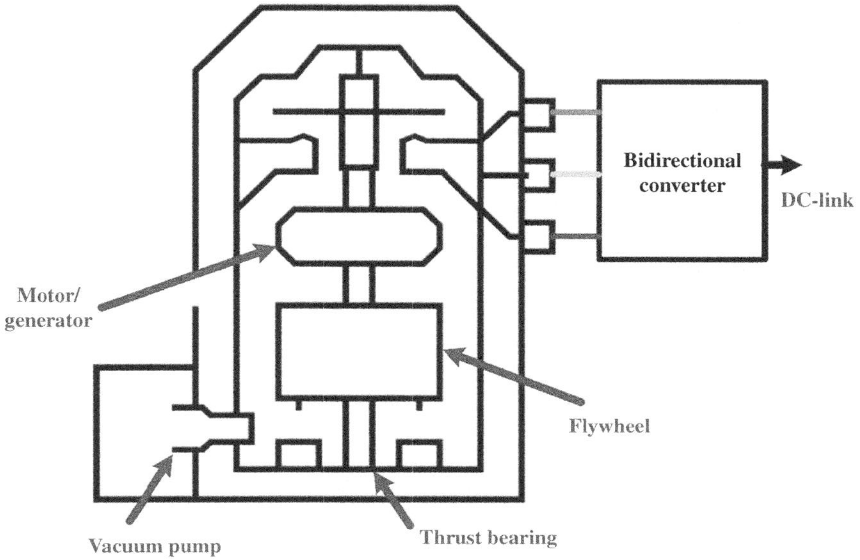

Figure 1.24 Schematic of the flywheel system.

for capturing and releasing energy quickly. Unlike conventional batteries, spring energy storage relies on mechanical deformation to store energy, which can be rapidly discharged to assist in acceleration, regenerative braking, or short bursts of power without chemical degradation. Recent research has explored compact, high-strength composite springs integrated into electric scooters, micro-mobility platforms, and auxiliary systems in larger EVs, showing improved efficiency in energy recapture and reduced load on the main battery. One of the key advantages is durability: spring systems can endure millions of cycles with minimal wear, making them ideal for frequent start-stop applications in urban driving. Additionally, hybrid approaches that combine springs with batteries or supercapacitors are being investigated to smooth peak power demands while supporting overall vehicle range. These approaches offer a promising path toward lighter, more resilient, and efficient energy-storage solutions in next-generation electric mobility [35–37].

1.3.6 Thermal Energy Storage

EVs use thermal storage mainly for managing battery and cabin temperatures. Battery packs employ thermal management systems (liquid cooling, heat pipes, etc.) to keep cells in a safe temperature range of 20–40 °C. Passive thermal storage in EVs often involves phase-change materials (PCMs) integrated into modules.

The latent heat of fusion in a PCM absorbs spikes in battery heat, flattening temperature profiles without active cooling. For example, some EV battery designs include wax or salt-based PCMs that melt at around 40–60 °C, storing heat during heavy discharge and solidifying during idle periods. This passive latent storage can improve safety and efficiency by stabilizing cell temperatures. Sensible heat storage (e.g., water jackets or coolant loops) is limited by low energy density (water stores only approximately 4.2 kJ/kg K); therefore, EV thermal systems rely mostly on circulation rather than bulk heat storage. In cold climates, waste heat from the battery or power electronics can be stored in small heat exchangers to preheat the cabin. Overall, EVs prioritize lightweight, compact thermal management (often with heat pumps) over large-scale thermal storage. In summary, EV thermal storage focuses on heat regulation for performance and safety (e.g., using PCMs and cooling loops) rather than long-term seasonal or grid-level heat storage [37].

1.4 Hybrid Energy-storage Systems

Hybrid energy-storage systems (HESS) combine two (or more) storage technologies to exploit their complementary strengths. A common pairing is a high-energy battery with a high-power supercapacitor. In such hybrids, the battery provides long-duration energy, whereas the supercapacitor oversees rapid power demands and absorbs transient spikes [38, 39]. For example, in an EV powertrain, a Li-ion battery is paralleled with an ultracapacitor (high power). During regenerative braking, the ultracapacitor soaks up the surge of energy, thereby smoothing the load on the battery. During acceleration or peak-load events, the capacitor can deliver a burst of power that the battery alone might struggle to supply. This arrangement protects the battery from extreme currents and can extend its usable life. Advanced energy management strategies ranging from simple rule-based logic to fuzzy controllers or model-predictive control decide how to split the power between the two devices. Typically, a bidirectional DC–DC converter links the battery and supercapacitor, regulating current flow in both directions as shown in Figure 1.25.

Careful design and control of HESS can significantly enhance performance. One recent study found that an optimally designed Li-ion/ultracapacitor hybrid in an EV extended the battery's useful life by roughly 47% and reduced its capacity degradation by about 23% (versus using the battery alone). Hybrid storage has been implemented in various applications beyond cars. For instance, rail transit systems have used battery–capacitor hybrids to handle frequent braking and door cycles, and wind turbines have paired flywheels with batteries for better power smoothing. Toyota has experimented with ultracapacitor-equipped vehicles (e.g., the Toyota Yaris Hybrid R concept used a supercapacitor to improve response).

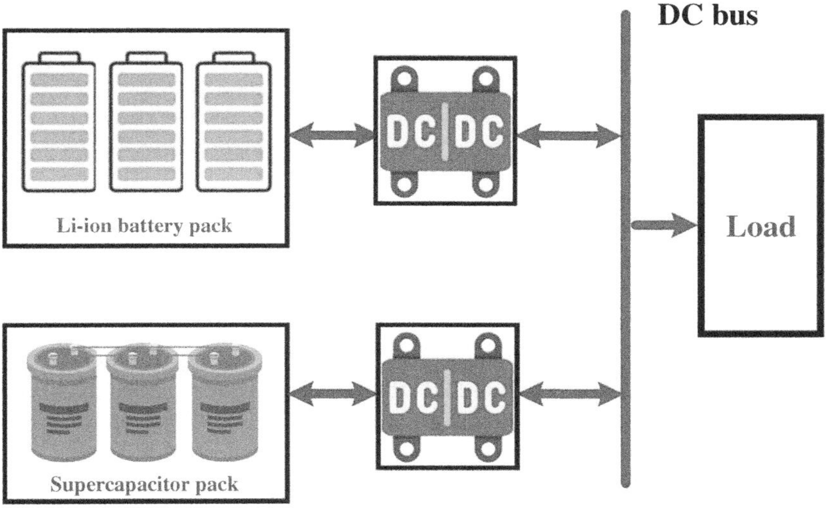

Figure 1.25 Li-ion battery-supercapacitor HESS configuration.

In all HESS designs, the challenge is managing two power sources: dedicated power electronics (e.g., bidirectional DC–DC converters) control the current between them, and an energy-management system decides when each element is used.

The benefits include improved system efficiency and longer lifetime (the supercapacitor shields the battery from large charge/discharge spikes), but these systems also add cost and complexity (extra components and control). To aid their development, specialized design and simulation toolboxes for hybrid energy-storage systems are now emerging, helping engineers to improve the sizing and control of these systems.

1.5 Comparative Analysis and Applications

Different energy-storage technologies tend to excel in different application domains:

- **Passenger EVs:** Modern EVs overwhelmingly use Li-ion batteries for propulsion. Typical packs today achieve approximately 200–250 Wh/kg at the pack level (depending on chemistry). Automakers are pushing next-generation chemistries (solid-state Li–metal, Li–S, high-Ni cathodes) to reach 400–500 Wh/kg or more, enabling much longer range. For short-term energy surges (e.g., rapid acceleration or stop-start operation), some vehicles incorporate ultracapacitors or flywheel auxiliaries. These devices store

Figure 1.26 Electric and hydrogen-based EVs.

high power but low energy. Ultracapacitors can recapture braking spikes, and small flywheels (or mechanical springs in shocks) have been tested to harvest suspension or brake energy. Such hybrid systems extend battery life by shaving power peaks. Fuel-cell (hydrogen) vehicles offer the potential for longer range but are still rare. By mid-2021, there were only approximately 40,000 hydrogen fuel-cell vehicles worldwide (out of over 1.4 billion total vehicles illustrated in Figure 1.26), i.e., well under 0.1% of the fleet [1–4].

- **Commercial and Heavy Transport:** Electric buses, trucks, and rail systems also rely on Li-ion batteries, often with larger total capacity. Durability and fast charging are emphasized in these applications. Some hybrid locomotives and buses use flywheel modules for regenerative braking because of their long cycle life. For example, a diesel–hybrid train may use a flywheel system with a capacity exceeding 100+ kWh to buffer acceleration loads. In warehouse equipment, both lead–acid and Li-ion batteries are commonly used. In the aerospace and defence sectors, where weight is prioritized, research vehicles use high-energy chemistries (Li–S, Li–polymer) and sometimes ultracapacitors for pulsed-power applications [37].
- **Vehicle-to-Grid (V2G) and Grid Services:** Aggregated EVs can function as distributed storage for the grid. With bidirectional chargers, EV batteries can be charged when renewable energy is abundant and can be discharged during peak demand. Studies have shown that fleets of charging EVs can smooth grid loads and provide frequency regulation. This emerging application treats parked EVs as a "virtual" power plant. In smart-grid pilots, AI and real-time control coordinate EV charging schedules, using both batteries and vehicle auxiliary storage (e.g., heat stored in thermal systems) to optimize energy flow. V2G integration is increasingly seen as a key use case for EV energy storage [37].

1.6 Challenges and Future Directions

Although energy-storage technologies have advanced rapidly, major challenges stay for widespread deployment and scaling:

- **Cost and Materials:** Battery packs remain a major cost driver for EVs, though prices have fallen (approximately \$100–150/kWh in mid-2020s). The need for long-duration storage (e.g., multi-hour reserves) is still costly if based on batteries. Meanwhile, critical minerals (Li, Co, and Ni) have supply constraints. Mining must ramp up to meet EV demand, and reducing reliance on scarce elements is vital. This motivates R&D on low-cost chemistries (e.g., Na-ion, iron-based) and better recycling [37].
- **Safety:** High-energy batteries pose fire risks if abused. Li-ion cells can experience thermal runaway at elevated temperatures or due to internal shorts. Effective thermal management and flame-retardant materials are essential. Similarly, mechanical systems (flywheels) must have containment to safely arrest rotor failures. Ongoing research in solid electrolytes aims to cut flammable liquids altogether.
- **Lifecycle and Degradation:** EV batteries degrade over time. Typical modern batteries may achieve 500–1000 full cycles before falling below approximately 80% capacity. After this, a battery is often unsuitable for automotive use, though as one review notes, spent EV batteries keep significant capacity for secondary use. Enhancing longevity is a focus: advanced BMS, gentler charging profiles, and hybrid systems (battery + capacitor) can extend cycle life. Flow and gravity systems effectively have unlimited cycles but are impractical in vehicles.
- **Recycling and Sustainability:** With millions of EV batteries reaching end-of-life, efficient recycling is critical. Only approximately 5–10% of Li-ion batteries were recycled by 2020, but efforts are accelerating. Current industrial processes (pyrometallurgy, hydrometallurgy) can recover metals but are energy-intensive. Newer "direct regeneration" methods aim to restore cathode materials for reuse, improving circularity. Policy and technology are also pushing digital tracking (battery passports) and automated disassembly to streamline EV battery recycling. Sustainable mobility will require designing batteries for ease of recycling and reusing second-life batteries in stationary storage to offset raw material needs [40].
- **Infrastructure and Scalability:** Deploying EVs at scale hinges on charger availability and grid upgrades. IEA reports over 5 million public chargers exist (doubling since 2022), but more fast-charging stations are needed to meet demand. Standardization (common charging protocols) and smart networks are being developed to avoid grid strain. Likewise, geographic limits mean some storage technologies cannot easily expand. Abundant-material systems (Na-ion, flow batteries using iron/organic materials) are promising for future scale. Ultimately, coordinating EV growth with raw-material supply and charging infrastructure is a major challenge ahead [37–40].

1.7 Emerging Technologies: Looking Forward

Research is pushing the boundaries of storage in several directions:

- **Solid-State Batteries:** Replacing liquid electrolytes with solid ion conductors promises higher energy density and enhanced safety. Many companies and research groups are developing solid-state Li–metal cells. These could potentially reach cell-level energy densities around 400–500 Wh/kg or more while eliminating flammable liquid electrolytes [32, 33]. The basic configuration of a solid-state battery is shown in Figure 1.27.
- **Lithium–Air (Li–O_2) and Other Metal–Air Batteries:** Lithium–air batteries have a very high theoretical specific energy (approximately 1000–1400 Wh/kg depending on design). This is because the cathode uses oxygen from air as the reaction partner, storing only Li and a porous carbon matrix. In practice, Li–air cells have struggled with cycle life, efficiency, and dealing with contaminants (moisture, CO_2) in the air. Recent research has achieved modest cycling (a few dozen cycles) and improved catalysts, but commercialization remains decades away. Other metal–air systems (zinc–air, aluminum–air) are also being studied for niche applications (e.g., long-endurance drones, backup), but none are near parity with Li-ion in portability yet [31–33].
- **Abundant Metal–Ion Batteries (Na^+, K^+, Mg^{2+}, etc.):** Na-ion batteries are rapidly maturing. Na-ion cells have lower energy density than Li-ion but use extremely common materials (salt) and cheaper cathode/anode materials. China has already commercialized Na-ion packs, and research suggests Na-ion can be competitive in cost. One analysis projects that Na-ion pack costs could fall to approximately $40–50/kWh at scale, far below current Li-ion prices. Importantly, Na is about 1000 times more abundant than Li. Potassium-ion and magnesium-ion chemistries are still at earlier R&D stages. Potassium's abundance is like Na's, and K-ion batteries show Li-ion-like behavior but with larger ions. Magnesium (Mg^{2+}) has the appeal of carrying 2 electrons per ion (theoretically higher volumetric energy), but practical Mg batteries have lower voltage and are slower. Advances continue, for example, new electrode materials and electrolytes are improving Na-ion performance (e.g., better cathodes, hard carbon anodes). Some analysts project that future

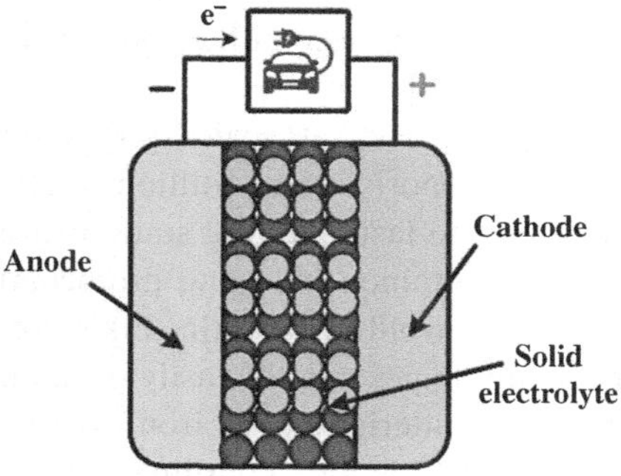

Figure 1.27 Solid-state battery configuration.

Na-ion systems might reach extremely low costs (e.g., \$30/kWh by mid-century) due to the cheap raw materials.

- **Next-generation Li Chemistries:** Beyond Na and K, researchers are exploring variants of Li batteries. Lithium–sulfur (Li–S) batteries use S cathodes to potentially reach approximately 500–600 Wh/kg at low cost. (S is very abundant and cheap.) The main challenge is cycle life: Li–S historically suffered from rapid capacity loss, but new electrode designs and electrolytes are extending its life. Researchers also investigate Li-selenium and Li-silicon systems, and even more exotic ideas like Li-chlorine ("memory battery") cells. A recent prototype Li–Cl_2 battery claimed approximately 362 Wh/kg (at the cell level) by using molten chlorine, but many hurdles remain (operating temperatures, corrosion). These next-generation Li systems aim to push energy density beyond current Li-ion limits, though they often trade off cycle life.
- **Smart Controls and AI:** Finally, the intelligence around EV storage is evolving as shown in Figure 1.28. BMS has undergone a significant transformation in recent years, especially within e-mobility. What was once limited to basic monitoring and protection has evolved into a highly intelligent and interconnected system capable of actively managing battery performance, safety, and longevity. Modern BMS designs increasingly incorporate AI, machine learning, and cloud-based communication, forming what are commonly described as smart or cloud-enabled BMS platforms. One of the most notable developments is the shift toward health-oriented battery management. Rather than focusing only on immediate performance, current

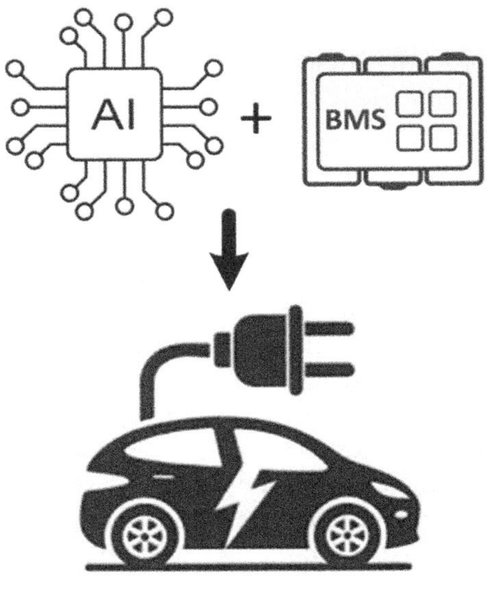

Figure 1.28 Configuration of AI-driven BMS for EVs.

strategies emphasize preserving long-term battery condition. This involves continuous estimation of parameters such as state of charge (SOC), state of health (SOH), state of power (SOP), and remaining useful life (RUL). By leveraging machine learning techniques, including neural networks and other data-driven models, these systems analyze large volumes of real-world operational data to identify early indicators of degradation, such as Li plating, thermal stress, and capacity loss. As a result, the BMS can intelligently regulate charging rates, depth of discharge, and thermal conditions, helping to extend battery life without noticeably affecting user experience.

The introduction of cloud-based BMS has further expanded these capabilities. In such systems, data collected from onboard sensors covering voltage, current, temperature, and impedance are transmitted to remote servers where more advanced computational models can be applied. This approach enables learning at the fleet level, where insights gained from large numbers of vehicles are used to refine predictive models and improve accuracy across the entire network. In addition, these systems support predictive maintenance and allow updates to be deployed remotely, which is particularly advantageous for commercial EV fleets where minimizing downtime is essential. Another area of rapid progress is intelligent charging management. AI-based algorithms can anticipate driving patterns, electricity pricing trends, and the availability of renewable energy sources such as solar and wind. Using these forecasts, the BMS can schedule charging during periods that are both cost-effective and environmentally favorable. Advanced methods, including reinforcement learning, enable real-time optimization of charging decisions while considering user needs such as departure time and required driving range alongside grid constraints. These capabilities are also critical for vehicle-to-grid applications, where EVs function as distributed energy assets. Thermal regulation has similarly benefited from these advancements. Since battery temperature plays a crucial role in performance, safety, and aging, modern systems integrate predictive thermal models with real-time sensor feedback to manage heating and cooling more effectively. Some approaches now employ digital twin models, which replicate the battery system virtually and simulate its behavior under varying conditions. This allows the system to expect and mitigate potential issues before they arise [41–44].

Vehicle-to-grid (V2G) technology is transforming EVs from mere consumers of electricity into active participants in the energy ecosystem. By enabling EVs to discharge stored energy back into the grid during peak demand or periods of renewable intermittency, V2G helps balance supply and demand while reducing strain on power plants. Recent developments have focused on integrating smart software platforms that coordinate fleets of vehicles, allowing them to respond in real time to grid signals and electricity pricing. This capability not only enhances

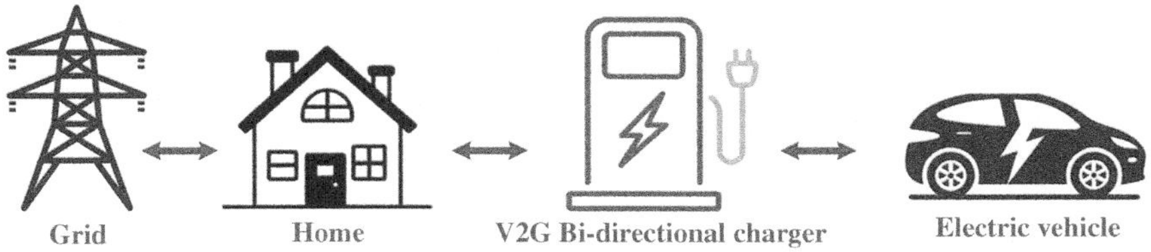

Figure 1.29 V2G configuration.

grid stability but also creates opportunities for owners to earn revenue through energy arbitrage or peer-to-peer trading. Essentially, EVs equipped with V2G act as mobile, distributed energy-storage units, forming the foundation of a more flexible, responsive, and decentralized electricity network (Figure 1.29).

In summary, energy storage has evolved from simple lead-acid batteries to a diverse suite of technologies that blend chemistry and systems engineering. Each technology has trade-offs; some excel at high power (ultracapacitors, flywheels), others at high energy (Li-ion, metal–air), and some at long life (flow batteries). Ongoing research is pushing these frontiers, from high-density solid-state cells to ultra-low-cost Na-ion. The goal is to integrate the right mix of storage solutions to achieve a reliable, affordable, and clean energy system. In practice, this will involve deploying proven technologies today while developing tomorrow's breakthroughs on solid-state, novel chemistries, and smart controls. As storage technologies continue to improve, guided by lessons from real-world pilots and underpinned by abundant new research, they will play a key role in enabling sustainable electric transportation and resilient power grids for the future.

References

1 International Energy Agency, "Electric vehicles." (Online). Available: https://www.iea.org/energy-system/transport/electric-vehicles. (Accessed: Apr. 2026).

2 Z. Alam and S. H. Gheewala, "Recharging the planet: The sustainability potential of second-life electric vehicle batteries," *J. Energy Storage*, vol. 144, p. 119720, 2026.

3 T. Sutikno, W. Arsadiando, and H. S. Purnama, "Battery types and recent developments for energy storage in electric vehicles: Technical criteria and battery management system," *Clean Energy*, vol. 9, no. 6, pp. 293–322, 2025.

4 A. Z. Wathoni, K. A. Madurani, C. W. Lai, and F. Kurniawan, "Comprehensive review of sodium-ion battery materials: Advances and performance challenges," *Chem. Phys. Mater.*, vol. 4, no. 4, pp. 344–359, 2025.

5 A. O. Ali, O. Abdelrehim, M. M. Saafan, M. R. Elmarghany, and A. M. Hamed, "Comprehensive review of battery management systems for electric vehicles: Thermal management, charging strategies, and emerging technologies," *J. Power Sources*, vol. 658, p. 238269, 2025.

6 International Energy Agency, "Global EV outlook 2025: Electric vehicle charging." (Online). Available: https://www.iea.org/reports/global-ev-outlook-2025/electric-vehicle-charging. (Accessed: Apr. 2026).

7 M. S. Eltohamy, M. H. Tawfiq, M.M. Ahmed, Z. Alaa, B. Mohammed, Ahmed I, Youssef H, Raouf A, "A comprehensive review of vehicle-to-grid (V2G) technology: Technical, economic, regulatory, and social perspectives," *Energy Convers. Manage.: X*, vol. 27, p. 101138, 2025.

8 X. Liu, W. Li, X. Guo, B. Su, S. Guo, Y. Jing, X. Zhang "Advancements in energy-storage technologies: A review of current developments and applications," *Sustainability*, vol. 17, no. 18, p. 8316, 2025.

9 V. Mahadevan, M. A. Rusho, and S. Yishak, "Critical review of energy storage systems: A comparative assessment of mechanisms, advantages, challenges, and integration with renewable energy," *Results Eng.*, vol. 26, p. 105589, 2025.

10 M. A. Rahman, J. H. Kim, and S. Hossain, "Recent advances of energy storage technologies for grid: A comprehensive review," *Energy Storage*, vol. 4, no. 6, p. e322, 2022.

11 D. A. Elalfy, E. Gouda, M. F. Kotb, V. Bureš, and B. E. Sedhom, "Comprehensive review of energy storage systems technologies, objectives, challenges, and future trends," *Energy Strategy Rev.*, vol. 54, p. 101482, 2024.

12 M. Costa, R. Maka, F. S. Marra, A. Palombo, and M. V. Prati, "Assessing techno-economic feasibility of cogeneration and power-to-hydrogen plants: A novel dynamic simulation model," *Energy Rep.*, vol. 10, pp. 1739–1752, 2023.

13 C. Chen, Y. Zhou, H. Fang, X. Peng, and L. Jiang, "Progress and challenges in energy storage and utilization via ammonia," *Surf. Sci. Technol.*, vol. 1, no. 1, p. 13, 2023.

14 J. Ren, H. Lou, N. Xu, F. Zeng, G. Pei, Z. Wang, "Methanation of CO/CO_2 for power-to-methane process: Fundamentals, status, and perspectives," *J. Energy Chem.*, vol. 80, pp. 182–206, 2023.

15 S. Biswas, A. P. Kulkarni, S. Giddey, and S. Bhattacharya, "A review on synthesis of methane as a pathway for renewable energy storage with a focus on solid oxide electrolytic cell-based processes," *Front. Energy Res.*, vol. 8, p. 570112, 2020.

16 M. Aziz, A. T. Wijayanta, and A. B. Nandiyanto, "Ammonia as effective hydrogen storage: A review on production, storage and utilization," *Energies*, vol. 13, no. 12, p. 3062, 2020.

17 P. Mehra, S. Saxena, and S. Bhullar, "A comprehensive analysis of supercapacitors and their equivalent circuits—A review," *World Electr. Veh. J.*, vol. 15, no. 8, p. 332, 2024.

18 Z. Li, H. Li, T. Qiu, F. Lin, and Y. Wang, "Reliability criterion of film capacitor based on moisture diffusion in encapsulation," *High Voltage*, vol. 8, no. 6, pp. 1242–1252, 2023.

19 X. Liu, S. Chen, Z. Xiong, K. Li, and Y. Zhang, "Tungsten oxide-based nanomaterials for supercapacitors: Mechanism, fabrication, characterization, multifunctionality, and electrochemical performance," *Prog. Mater. Sci.*, vol. 130, p. 100978, 2022.

20 K. Detka and K. Górecki, "Selected technologies of electrochemical energy storage—A review," *Energies*, vol. 16, no. 13, p. 5034, 2023.

21 I. Gwayi, S. P. Ayeng'o, and C. Z. Kimambo, "Selection of electrochemical and electrical energy storage systems for off-grid renewable energy mini-grids: A review," *Cleaner Eng. Technol.*, vol 25, p. 100906, 2025.

22 A. Clemente and R. Costa-Castelló, "Redox flow batteries: A literature review oriented to automatic control," *Energies*, vol. 13, no. 17, p. 4514, 2020.

23 G. G. Njema, R. B. Ouma, and J. K. Kibet, "A review on the recent advances in battery development and energy storage technologies," *J. Renewable Energy*, no. 1, p. 2329261, 2024.

24 A. Robert, B-J. R. Mungyeko Bisulandu, A. Ilinca, and D. R. Rousse, "Hybrid wind–redox flow battery system for decarbonizing off-grid mining operations," *Appl. Sci.*, vol. 15, no. 13, p. 7147, 2025.

25 A. Samanta and S. Williamson, (eds.), *Battery Technologies for E-Mobility: Innovations, Applications, and Advanced Management Systems*. Boca Raton, FL, USA: CRC Press, 2025.

26 C. J. Vijaykumar, S.S. Bulla, C. Chavan, H.P. Nagaswarupa, R.F. Bhajantri, B. Pushpa, R. Naik, "Comprehensive review of solid-state batteries: Theoretical conceptualization to sustainable energy storage technologies," *J. Energy Storage*, vol. 151, p. 120431, 2026.

27 G. J. May, A. Davidson, and B. Monahov, "Lead batteries for utility energy storage: A review," *J. Energy Storage*, vol. 15, pp. 145–157, 2018.

28 A. Poullikkas, "A comparative overview of large-scale battery systems for electricity storage," *Renew. Sustain. Energy Rev.*, vol. 27, pp. 778–788, 2013.

29 A. R. Dehghani-Sanij, E. Tharumalingam, M. B. Dusseault, and R. Fraser, "Study of energy storage systems and environmental challenges of batteries," *Renew. Sustain. Energy Rev.*, vol. 104, pp. 192–208, 2019.

30 M. M. Hasan R. Haque, M. I. Jahirul, M.G. Rasul, I. M. Fattah, N. M. Hassan, M. Mofijur, "Advancing energy storage: The future trajectory of lithium-ion battery technologies," *J. Energy Storage*, vol. 120, p. 116511, 2025.

31 S. Yari, A. C. Reis, Q. Pang, and M. Safari, "Performance benchmarking and analysis of lithium–sulfur batteries for next-generation cell design," *Nat. Commun.*, vol. 16, no. 1, p. 5473, 2025.

32 A. Singh et al., "Chemistry in rechargeable zinc–air battery: A mechanistic overview," *Catal. Today*, vol. 445, p. 115108, 2025.

33 B. B. Adetokun, O. Oghorada, and S. J. Abubakar, "Superconducting magnetic energy storage systems: Prospects and challenges for renewable energy applications," *J. Energy Storage*, vol. 55, p. 105663, 2022.

34 X. Li and A. Palazzolo, "A review of flywheel energy storage systems: State of the art and opportunities," *J. Energy Storage*, vol. 46, p. 103576, 2022.

35 S. Guo, L. Yang, Y. Yuan, Z. Zhang, and X. Cao, "Elastic energy storage technology using spiral spring devices and its applications: A review," *Energy Built Environ.*, vol. 4, no. 6, pp. 669–679, 2023.

36 F. Akkus and M. Z. Isik, "A review of thermal management systems of lithium-ion batteries used in electric vehicles," *J. Traffic Transp. Eng.*, vol. 12, no. 6, pp. 1763–1783 2025.

37 M. Waseem, G.S. Lakshmi, M. Ahmad, and M. Suhaib, "Energy storage technology and its impact in electric vehicle: Current progress and future outlook," *Next Energy*, vol. 6, p. 100202, 2025.

38 S. M. Benoy, M. Pandey, D. Bhattacharjya, and B. K. Saikia, "Recent trends in supercapacitor–battery hybrid energy storage devices based on carbon materials," *J. Energy Storage*, vol. 52, p. 104938, 2022.

39 B. Pang, H. Zhu, Y. Tong, and Z. Dong, "Optimal design and control of battery–ultracapacitor hybrid energy storage system for BEV operating at extreme temperatures," *J. Energy Storage*, vol. 101, p. 113963, 2024.

40 J. Chacana-Olivares, B. Peceño, M. Grageda, C. Cruz, and L. Rojas, "Lithium-ion battery recycling: A perspective on key challenges and opportunities," *npj Mater. Sustain.*, vol. 3, no. 1, p. 38, 2025.

41 P. Ranjbaran, J. Ebrahimi, A. Bakhshai, and P. K. Jain, "A comprehensive review of topologies and energy management methods for hybrid energy storage systems in electric vehicle applications," *IEEE Trans. Power Electron.*, vol. 41, no. 4, pp. 6149–6178, 2025.

42 A. Samanta, M. Sharma, W. Locke, and S. Williamson, "Cloud-enhanced battery management system architecture for real-time data visualization, decision making, and long-term storage," *IEEE J. Emerg. Sel. Topics Ind. Electron.*, vol. 6, no. 4, pp. 1700–17112025.

43 C. Chetri and S. Williamson, "Health-conscious fast charging for electrified aircraft batteries using a multistage-constant-current temperature-controlled strategy," *IEEE J. Emerg. Sel. Topics Power Electron.*, vol. 13, no. 4, pp. 4635–4642, 2025.

44 Y. Achour, A. El Mokrini, and A. Berrada, "Integration of gravity energy storage and deep learning for optimal vehicle-to-grid (V2G) systems in residential microgrids," *Energy Storage Saving*, In press, 2026.

2

Battery Fundamentals

2.1 Introduction

2.1.1 Historical Development of Battery Technology

The historical development of battery technology for electric vehicles (EVs) has evolved through distinct phases of innovation, transitioning from foundational chemistry to advanced materials Figure 2.1. Between 1976 and 1985, research focused on identifying battery systems and evaluating energy resources to meet the early EV requirements [1]. While many systems remained experimental, the lead-acid battery was established as a viable near-term solution due to its cost-effectiveness, alongside early explorations into lithium technologies driven by the 1970s energy crisis. From 1986 to 1995, the focus shifted toward alternative chemistries, such as sodium-nickel chloride and nonaqueous lithium designs, to overcome energy density limitations [2]. This era was heavily influenced by socio-political pressures and environmental regulations, such as the California ZEV mandate, which spurred the emergence of hybrid power systems and interactive software to model vehicle life-cycle costs [3].

The decade spanning 1996 to 2005 marked the commercialization of high-performance systems, with nickel-metal hydride (NiMH) and lithium-ion batteries becoming the leading technologies [4]. Innovations during this time included the development of scalable lithium polymer batteries and the introduction of intelligent battery management systems (BMS) for real-time state-of-charge calculation and thermal diagnostics. Between 2006 and 2015, research priorities expanded to include efficiency and environmental impact, leading to advancements in hydrogen fuel cells, nanostructured electrode materials [5] (such as silicon and graphene), and the economic evaluation of second-life applications like vehicle-to-grid (V2G) systems.

In the most recent phase from 2016 to 2025, the industry has pivoted toward sustainability and the circular economy [6]. Significant progress has been made

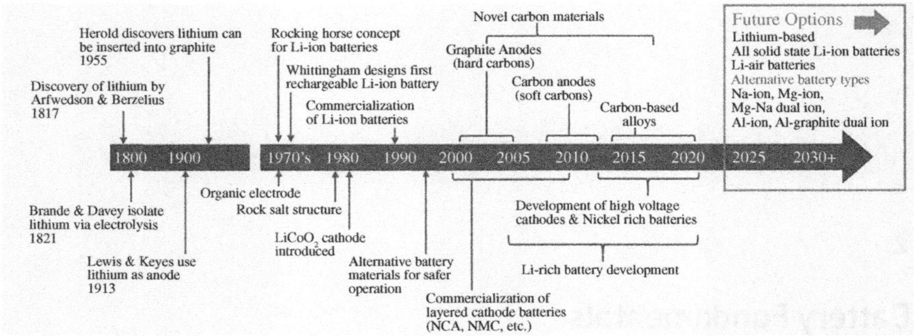

Figure 2.1 History of batteries.

in optimizing battery modules and packs through cell-to-pack technologies and advanced liquid cooling systems to prevent thermal runaway [7]. Current research emphasizes sophisticated recycling methods, including pyrometallurgy and hydrometallurgy to reclaim critical materials like cobalt and lithium [8]. Furthermore, the integration of digital twins and machine learning models has become a forward-looking standard, providing virtual replicas for real-time performance optimization and predictive analytics for the entire battery life cycle [9].

2.1.2 Historical Development of Automotive Propulsion for EVs
Early Development Stage (Late 1800–1930)

In its infancy, electric propulsion was the dominant automotive technology, well ahead of the internal combustion engine. During the late 1920s to early 1930s, EVs outnumbered gasoline vehicles by a ratio of 3:1 [10]. They were the primary tools for local transportation, having evolved directly from horse-drawn carriages. However, by 1930, gasoline vehicles had overtaken EVs due to the maturity of mass production, specifically the manufacturing revolution led by the Ford Model T [11], which made gasoline cars more affordable to the public. Furthermore, the lack of charging infrastructure, limited travel range, and the discovery of cheap, abundant petroleum led to the decline of early electric propulsion.

2.1.2.1 Mid-term Development (1930s–1980s)

Following the dominance of the combustion engine in 1935, EV production for personal transportation largely came to a halt. Interest only reignited during the 1960s and 1970s due to energy independence concerns surrounding OPEC and the 1970s energy crisis. This period was marked by government-mandated fuel efficiency standards and experimental test fleets, such as the U.S. Postal Service's order of 350 EVs. Despite these efforts, development quieted down again by the late 1980s due to limited performance and a continued lack of broad infrastructure support.

2.1.2.2 Modern and Future Development (Late 1990–2012)

The modern era began with a surge of fleet-based EVs in the late 1990s, including the GM EV1, Ford Ranger EV pickup, Toyota Rav4 EV, and Honda EV [12]. While most of these early modern models were eventually discontinued or recycled, they paved the way for the hybrid electric vehicle (HEV). The 1999 launch of the Honda Insight and the subsequent success of the Toyota Prius demonstrated the maturity and market potential of electrified propulsion. By 2012, the trend shifted back toward pure electrification and plug-in hybrids, supported by sophisticated system management, safer lithium battery technology, and government tax incentives.

2.2 Different Battery Types

The automotive industry has established lithium-ion batteries (LIBs) as the primary energy storage solution due to their superior specific capacity, lightweight design, and rapid charging capabilities compared with legacy lead-acid and NiMH technologies. By utilizing graphite anodes and lithium-ion electrolytes instead of unstable lithium metal, LIBs form a stable solid electrolyte interface (SEI) layer that significantly enhances safety and performance [13]. The technology's versatility is reflected in a diverse range of chemistries: lithium iron phosphate (LFP) is valued for thermal stability and cycle life, nickel manganese cobalt (NMC) provides a balance of energy density and stability, and lithium nickel cobalt aluminum oxide (NCA) supports extended ranges in high-performance vehicles [14]. Furthermore, specialized variants such as lithium titanate oxide (LTO) offer exceptional performance in extreme temperatures and fast-charging scenarios [15], whereas emerging solid-state batteries (SSBs) represent the next frontier in safety and energy density [16]. Ongoing research, such as the studies focused on 21700 and 18650 NMC and NCA cells from manufacturers such as Samsung SDI, LG Chem, and Molicel, continues to drive the advancement of these systems within the sustainable mobility landscape [17]. Table 2.1 shows a detailed comparison for all types of lithium-ion batteries available commercially.

2.3 Construction

2.3.1 Lithium-ion Battery

The construction of a lithium-ion battery (LIB) is a sophisticated multilayered assembly designed to facilitate the movement of lithium ions between two electrodes (Figure 2.2). At its core, the battery consists of a negative anode, typically made of graphite, and a positive cathode, which uses specialized chemistries such

Table 2.1 Comparison of new lithium-ion battery chemistry.

	NMC	LFP	NCA	LTO	LCO	SSB
Nominal voltage (V)	3.7	3.3	3.7	2.4	3.7	3
Operating voltage range (V)	3.0–4.2	2.5–3.65	2.5–4.2	1.8–2.85	3.0–4.2	2.5–4.2
Specific energy (Wh/kg)	140–180	80–130	200–300	150–240	50–80	260–400
Specific power (W/kg)	500–3000	1400–2400	500–1,500	3000–5100	200–400	374–500
Charge (C-rate)	0.7C–1C	1C typical	0.7C, charges to 4.20 V, fast charging possible with some cells	1C typical; 5C maximum, charges to 2.85 V	0.7–1C	1C
Discharge (C-rate)	1C–2C	1C typical, 25C on some cells	1C	10C possible, 30C 5s pulse	1C	1C
Cycle life	1000–2000 cycles	5000–7000, up to 12,000 possible in some cells	500	3000–7000	500–1000	10,000+
Safe operating temperature (°C)	−10 to 45	−20 to 50	−20 to 50	−30 to 55	−10 to 45	−30 to 150
Thermal runaway	210 °C (410 °F)	270 °C (518 °F)	150 °C (302 °F)	200 °C +	150 °C (302 °F)	
Safety	Prone to thermal runaway, flame, and explosion	Lowest rate of thermal runaway and self-ignition	Prone to thermal runaway, flame, and explosion	Low-temperature operation and charge	Dangerous chemistry Limited Lifespan	Thermally stable battery
Environmental impact	Cobalt and Nickel have a negative impact on the environment	Use safer material	Heavy metal extraction and improper disposal	Heavy metal extraction and improper disposal	Rare material: Cobalt and improper disposal	More environmentally friendly than other Li-ion batteries
Cost (USD/kWh)	151–177	118–152	156–172	220–240	135–150	80–90

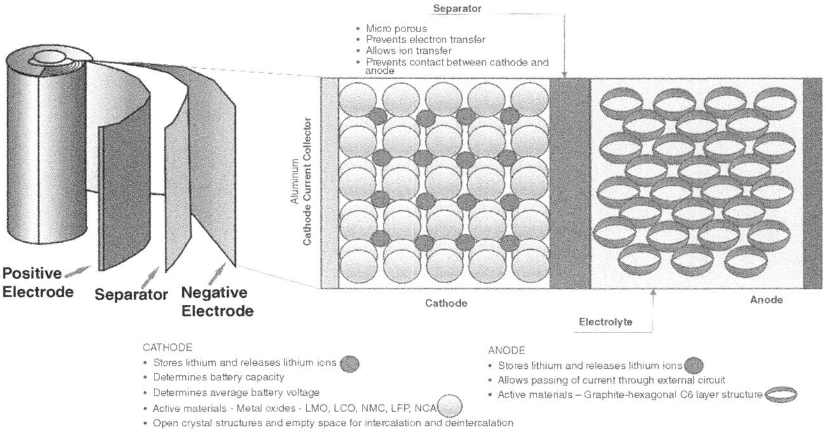

Figure 2.2 Internal structures of lithium-ion battery.

as NMC or NCA to define its energy density and performance [18]. These electrodes are physically separated by a microporous membrane that prevents short internal circuits while allowing ions to travel through a liquid electrolyte composed of lithium-ion salts. This internal configuration promotes the formation of a stable solid electrolyte interface (SEI) layer on the anode, which is critical for long-term safety and preventing dangerous lithium plating [19]. In automotive applications, these cells are manufactured in standardized form factors, such as the 18650 or 21700 cylindrical formats.

To meet the high power and energy demands of EVs, individual cells are integrated into a larger hierarchy of modules and packs. This structural assembly includes advanced mechanical components for stability and a comprehensive Thermal Management System often utilizing liquid cooling channels or phase-change materials to maintain optimal operating temperatures and prevent thermal runaway [20]. Central to this construction is the intelligent BMS, which monitors individual cell voltages and temperatures in real-time to ensure balanced charging and overall system reliability [21]. More recent structural innovations, such as cell-to-pack and cell-to-chassis technologies, aim to increase volumetric efficiency by eliminating intermediate modules and integrating cells directly into the vehicle structure [22].

2.4 Operating Principles

The operating principle of a lithium-ion battery centers on the reversible movement of lithium ions between two electrodes, a process known as intercalation. During discharge, when the battery is powering a device, a chemical

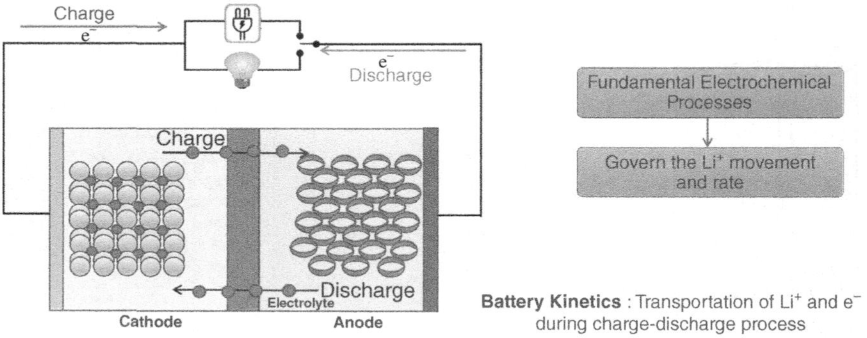

Figure 2.3 Battery working and kinetics.

oxidation reaction occurs at the anode (negative electrode), releasing electrons into the external circuit and positive lithium ions into the internal electrolyte [23]. These ions migrate through the electrolyte and a microporous separator to reach the cathode (positive electrode), where they are inserted into the crystal lattice of the cathode material (Figure 2.3). To maintain electrical neutrality, the electrons travel through the external device to meet the ions at the cathode, thereby creating the electric current that provides power [24].

Conversely, during charging, the process is reversed by an external power source. The applied voltage forces lithium ions out of the cathode's molecular structure and drives them back through the electrolyte and separator into the anode, where they are stored between layers of graphite [25]. The process of ion shuttling is highly efficient because it does not involve the physical breakdown of the electrodes, allowing the battery to be recharged thousands of times [26]. The separator plays a critical safety role during this entire cycle by acting as a physical barrier that prevents the two electrodes from touching and causing a short circuit, while its microscopic pores remain large enough to allow the lithium ions to pass freely [27].

2.5 Internal Kinetics

The internal kinetics of a lithium-ion battery refer to the rates and mechanisms that govern how quickly lithium ions and electrons move during operation. While thermodynamics determines the battery's theoretical voltage, kinetics determines its actual performance, such as the amount of power it can deliver and how fast it can charge [28]. These kinetics are primarily limited by three barriers:

charge transfer at the electrode surfaces, diffusion within the solid electrode particles, and ionic transport through the electrolyte [29].

At the interface where the solid electrode meets the liquid electrolyte, the Butler–Volmer kinetics describe the charge transfer reaction [30]. This is the energy barrier an ion must overcome to leave the electrolyte and enter the solid electrode (or vice versa). Once inside the electrode, the ions move via solid-state diffusion, a relatively slow process governed by Fick's Laws, where ions hop between vacant sites in the crystal lattice. In 2025, many high-performance batteries use nano-structured materials to shorten these diffusion paths, significantly boosting power density [31]. Finally, the solid-electrolyte interphase (SEI), a thin passivation layer on the anode, acts as a kinetic gatekeeper; it must be conductive enough to let lithium ions through but resistive enough to block electrons and prevent further electrolyte breakdown.

2.6 Charging and Discharging Characteristics

The charging and discharging characteristics of lithium-ion batteries are defined by specific voltage and current profiles that ensure both efficiency and safety. Charging typically follows a two-stage protocol known as constant current constant voltage (CC-CV) [32]. In the first stage (CC), the charger provides a steady current, causing the battery voltage to rise rapidly until it reaches its upper limit, usually around 4.2 V for a standard cell [33]. This phase efficiently restores about 80% of the capacity. Once the limit is reached, the system switches to the CV stage, where the voltage is held constant while the current gradually tapers off until the battery is fully saturated. This prevents overcharging and chemical stress [34].

Discharging characteristics are characterized by a distinctive S-shaped curve that represents the relationship between voltage and the state of charge (SoC) [35]. When a load is applied, the voltage initially experiences a sharp IR drop due to internal resistance [36], followed by a long, relatively flat plateau where the voltage remains stable for most of the cycle (typically between 3.7 and 3.3 V). This stability is a key advantage, as it allows electronic devices to operate at a consistent power level. As the battery nears depletion, the voltage enters a knee and drops precipitously. If the voltage falls below a critical cut-off threshold (often 2.5–3.0V), the BMS disconnects the load to prevent permanent copper dissolution and internal damage. These characteristics are highly sensitive to the C-rate (speed of discharge) and temperature higher currents, or extreme cold will cause the voltage to sag more significantly and reduce the usable capacity.

2.7 Battery Management System

2.7.1 Battery Monitoring and State Estimation

The BMS is a crucial component in the battery pack of an EV, ensuring its safety, efficiency, and longevity. The BMS is responsible for monitoring and managing the performance of individual battery cells and the entire battery pack under varying operational conditions [37]. Its importance lies in several key functions.

2.7.2 Safety Management

The BMS prevents hazardous situations by monitoring the temperature, voltage, and current of each cell. It ensures that the battery operates within safe limits, protecting against overcharging, over-discharging, thermal runaway, and short circuits, which could lead to fires or battery failure.

2.7.3 Battery Performance Optimization

By balancing the charge across individual cells in the battery pack, the BMS ensures uniform performance. This cell balancing prevents any single cell from degrading faster than others, enhancing the overall efficiency and reliability of the pack [38].

2.7.4 Longevity Enhancement

The BMS prolongs battery life by avoiding stress conditions such as deep discharges or high charge currents. It manages the charge–discharge cycles optimally, reducing the wear and tear on battery components over time.

2.7.5 Thermal Management

The BMS works with the EV cooling or heating system to maintain the battery within an optimal temperature range. This is critical for preserving battery performance and preventing failures in extreme temperatures.

2.7.6 Energy Efficiency

By managing the charging and discharging processes efficiently, the BMS reduces energy losses and improves the overall energy efficiency of the EV, contributing to a longer driving range per charge.

2.7.7 Communication and Control

The BMS communicates with other EV systems, such as the powertrain and charger, ensuring seamless integration and operation. It also enables diagnostics and fault detection, alerting users and maintenance teams to potential issues before they become critical.

2.7.8 Architecture of LIBs with BMS

LIB packs in EVs are typically assembled by connecting individual cells in series and parallel configurations to achieve the desired voltage and capacity. These cells are grouped into modules, which are further connected to a slave BMS for monitoring and control. The slave BMS in each module measures parameters like voltage, current, and temperature, ensuring the safety and balancing of the cells. Data from the slave BMS modules are transmitted to the master BMS through daisy-chain communication or transformer physical layer (TPL) communication protocols, enabling centralized monitoring and efficient management of the entire battery pack [39]. Figure 2.4 shows the relation among modules, master, and slave BMS architecture.

2.7.9 Protection in BMS

A BMS incorporates critical protection features to ensure the safe operation of LIB packs. Overcharge protection prevents the cells from exceeding their maximum voltage, which could lead to overheating, capacity loss, or even thermal runaway. Similarly, over-discharge protection safeguards the cells from being depleted below their minimum voltage, avoiding irreversible damage and reduced lifespan. The BMS also includes starting surge protection to limit inrush currents during power-up, which can otherwise damage the cells and connected components. These features collectively enhance the reliability, safety, and longevity of the battery pack [40]. Figure 2.5 shows the protection circuit associated with BMS architecture.

2.7.10 Communication in BMS

The Controller Area Network (CAN) bus has become a widely adopted communication protocol in various industries, particularly in automotive, industrial, and in BMS due to its robustness, fault tolerance, and real-time data transfer capabilities. Initially developed by Bosch in the 1980s for automotive applications, CAN bus offers several advantages such as high-speed communication, error detection,

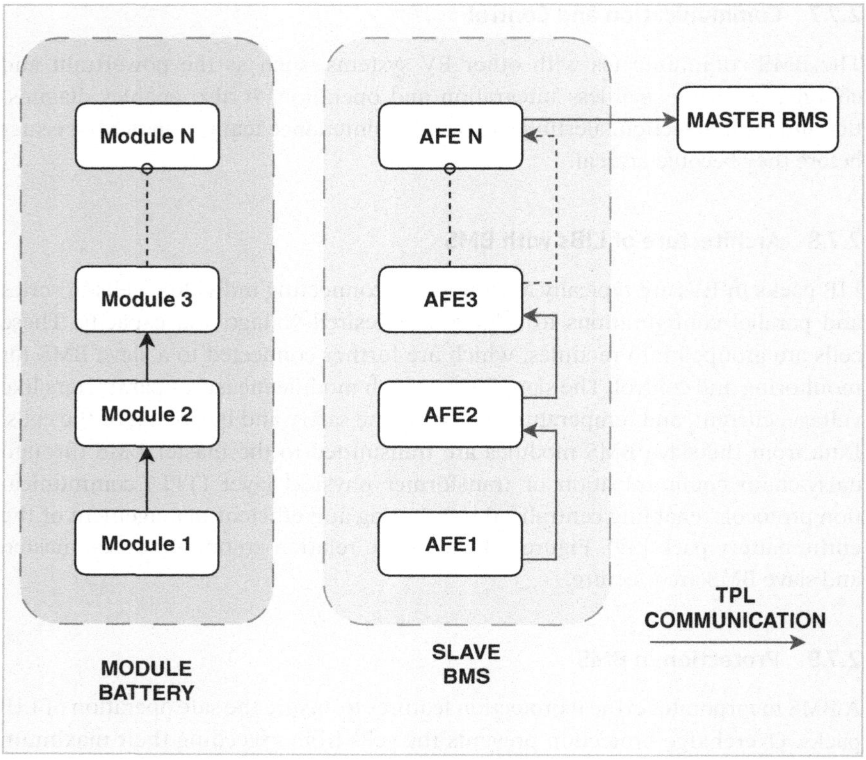

Figure 2.4 Battery management system architecture.

and the ability to handle complex networks of devices with minimal wiring. These features make it ideal for environments with high electromagnetic interference and where real-time performance is critical, such as in BMS.

In the context of BMS, CAN bus communication plays a pivotal role in monitoring and controlling the parameters of LIB packs. Several studies highlight the integration of CAN bus in BMS for efficient data exchange among battery cells, temperature sensors, and central controllers. For instance, CAN bus allows the BMS to communicate critical information about battery voltage, current, SOC, temperature, and SOH in real time, which enables the system to prevent overcharging, deep discharging, and thermal runaway, thereby enhancing the safety and longevity of the battery. The use of CAN bus in BMS allows for precise monitoring of individual cells, which is essential for balancing the cells and optimizing the overall battery performance.

Moreover, the decentralized nature of CAN bus supports the integration of multiple devices in a modular BMS architecture. Each battery module or cell

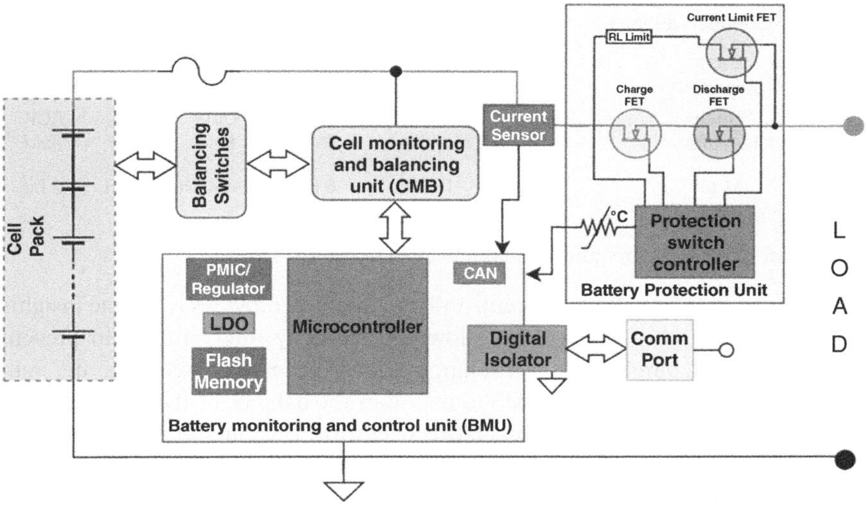

Figure 2.5 BMS architecture.

communicates with the central controller via CAN, allowing for scalable and flexible system design. CAN bus communication enables the real-time data acquisition and control necessary for optimal battery management in EVs and renewable energy systems. The ability to prioritize and manage communication between multiple devices over a single bus line reduces the complexity of wiring, leading to cost savings and enhanced system reliability.

The fault tolerance and noise immunity of CAN bus also make it well-suited for automotive applications, where electrical noise and physical stress are common. However, CAN's error detection mechanisms, including cyclic redundancy checks (CRC) and acknowledgement signals, ensure that corrupted data is retransmitted, further enhancing the reliability of the communication. Additionally, the protocol's support for diagnostics, remote monitoring, and firmware updates through a CAN interface facilitates system maintenance and performance optimization, a complete CAN architecture is shown in Figure 2.6. CAN communication is widely used in the automotive sector and offers various efficient features, such as cyclic redundancy and secured data transfer. Therefore, the proposed method uses CAN for data transmitting and collecting, which are later used for various estimation techniques, such as SOC and surface temperature [41].

2.7.11 Data Monitoring

Data monitoring in BMS is essential for optimizing battery performance, ensuring safety, and prolonging the lifespan of LIB packs. Continuous monitoring of key

	Extended arbitration							
	Base arbitration							
SOP	Base identifier	SRE	IDE	Extended identifier	Control field	Data field	CRC field	ACK field
1 bit	11 bit	1 bit	1 bit	18 bit	6 bit	0–64 bit	16 bit	2 bit

Figure 2.6 CAN Bus communication protocol architecture.

parameters, such as voltage, current, and temperature, provides real-time insights into the condition of the battery, allowing for timely interventions to prevent issues like overcharging, deep discharging, or overheating. Accurate and reliable data monitoring enables BMS to maintain the balance of the cells, improve energy efficiency, and reduce degradation over time. Data monitoring also plays a pivotal role in the thermal management of the battery, as temperature fluctuations can significantly affect performance and safety. The importance of monitoring temperature at the cell and module levels, with the BMS controlling thermal balancing to avoid thermal runaway or hotspots.

Furthermore, data monitoring facilitates predictive maintenance by continuously tracking the SOH and degradation patterns of the battery. By analyzing trends in key performance indicators, the BMS can forecast the need for maintenance or replacement of individual cells, helping to optimize battery life and reduce operational costs. The application of machine learning algorithms in BMS for predictive maintenance, where historical data on voltage, current, and temperature are used to model the expected performance of the battery and predict failures before they occur. This proactive approach helps prevent costly repairs and downtime in critical applications like EVs and energy storage systems.

Data monitoring also enhances the integration of the BMS with external systems, such as charging stations, powertrains, and grid systems, especially in vehicle-to-grid (V2G) applications. Real-time data transfer and communication protocols, such as CAN bus, enable the BMS to share information with external controllers for optimal energy flow and synchronization with the grid. This level of integration not only improves the efficiency of the overall system but also allows for remote monitoring and diagnostics, reducing the need for manual inspections and enabling fast response times in case of failures.

2.7.12 CAN Bus Data Decoder

The development of CAN data decoders for BMS plays a vital role in interpreting and managing the complex data transmitted over the CAN bus in modern LIB systems. CAN bus communication enables real-time, bidirectional data transfer between various components of the BMS, such as the central controller, battery cells, temperature sensors, and external systems. The data transmitted is often in raw hexadecimal or binary format, requiring a decoding process to convert it into meaningful parameters like voltage, current, temperature, SOC, and SOH. The role of CAN data decoders is crucial in ensuring that the data received from different modules is properly processed and mapped to specific functions, such as cell balancing, charging, and fault detection.

The importance of a reliable CAN data decoder is essential for accurate and efficient communication between the BMS and external devices, such as monitoring systems and vehicle powertrains. These decoders help to interpret complex CAN messages and convert them into human-readable values, allowing engineers to visualize key metrics like voltage levels, current flow, and temperature profiles. The decoder must also account for different communication formats and protocols used by various battery management modules, as these can vary between manufacturers and system configurations. For example, some systems use standard protocols such as the J1939 or ISO 11898, while others may have proprietary formats, requiring custom decoding algorithms to ensure compatibility and accuracy [42].

The efficiency of CAN data decoders is further enhanced using software tools and libraries. The implementation of Python-based decoders for CAN data, where libraries such as python-can and CANalyzer are used to interface with the CAN bus and decode messages in real time. These tools allow for easy integration with data visualization platforms, such as Grafana or MATLAB, enabling live monitoring and diagnostics of the battery system's performance. Moreover, CAN data decoders also enable the detection of anomalies or faults in the system by interpreting error flags, which can trigger warnings or shutdown protocols to prevent damage to the battery.

2.8 Introduction to Configurations and System Design Consideration

System design and configuration are the blueprint stages where individual lithium-ion cells are transformed into a functional power source for a specific application. At the core of configuration is the decision between series and parallel connections. Wiring cells in series connect the positive terminal of one to the negative of the next, which adds their individual voltages together while keeping the capacity the same for instance, four 3.7 V cells in series create a 14.8 V pack. However, parallel connections join all positive terminals together and all negative terminals together, which keeps the voltage constant but adds to the amp-hour capacities, allowing for longer runtimes. Most large-scale systems, like those in EVs, utilize a series-parallel (S-P) hybrid configuration to achieve both high voltage for propulsion efficiency and high capacity for range [43].

Beyond electrical wiring, system design considerations must address mechanical, thermal, and electronic integration to ensure reliability. Thermal management is a primary design pillar, as lithium-ion batteries perform optimally within a narrow temperature window (typically 15 °C to 35 °C) engineers must choose between passive air cooling, active fan systems, or liquid cooling plates to prevent "hot spots" that cause uneven aging or thermal runaway. Mechanical design involves creating a robust enclosure that protects cells from vibration, moisture, and impact while providing structural integrity. Finally, the battery management system (BMS) must be integrated as the electronic control layer, managing cell balancing to ensure that one "weak" cell doesn't compromise the entire string's performance. In 2025, these designs often focus on cell-to-pack (CTP) technology, which eliminates intermediate module housings to save weight and increase the overall energy density of the system.

2.9 Battery Degradation

Battery degradation is an inevitable process where a cell gradually loses its ability to store and deliver energy, manifesting primarily through capacity fade and power fade (increased internal resistance). This decline is driven by complex chemical and mechanical transformations, most notably the continuous growth of the solid electrolyte interphase (SEI) on the anode. While a thin SEI layer is necessary to protect the electrode, repeated cycling the movement of lithium ions back and forth causes the layer to thicken, trapping active lithium that can no longer participate in energy storage. Mechanical stress also plays a significant role; as lithium ions enter and exit electrode particles, the material physically expands and contracts, which can lead to microscopic particle cracking. These cracks expose fresh electrode surfaces, triggering further electrolyte decomposition and SEI formation, creating a snowball effect of degradation [44].

External factors significantly influence the rate of these internal kinetics. High temperatures act as a catalyst, accelerating the side reactions that consume the electrolyte and dissolve transition metals from the cathode, which can then migrate and poison the anode. Conversely, low temperatures or high-speed charging can lead to lithium plating, where ions deposit as metallic lithium on the surface of the anode rather than intercalating into it, this not only causes permanent capacity loss but can also form needle-like dendrites that risk puncturing the separator. Additionally, calendar aging occurs even when the battery is idle, driven primarily by the state of charge (SoC), keeping a battery at a constant 100% charge increases the internal voltage stress, speeding up chemical breakdown compared to storing it at a mid-range capacity.

2.10 Characteristics and Performance of Lithium-ion Batteries

The performance of lithium-ion batteries is defined by a unique combination of high-energy density, exceptional round-trip efficiency, and long cycle life, which collectively make them the dominant choice for modern electronics and EVs. As of 2025, commercial cells typically offer energy densities between 150 and 250 Wh/kg, allowing for lightweight designs that do not sacrifice runtime, while their high nominal voltage (typically 3.6–3.7 V) reduces the number of cells needed to achieve high-voltage targets. These batteries are remarkably efficient, boasting 90–95% round-trip efficiency, meaning very little energy is lost as heat during the charge–discharge process [45]. Therefore, their low self-discharge rate is often less than 2% per month ensures that devices remain ready for use even after long periods of storage, a significant improvement over older nickel-based chemistries.

Performance is also measured by the C-rate, which indicates how quickly a battery can be charged or discharged relative to its total capacity, for example, a 1C rate means a full discharge in one hour. Modern lithium-ion systems can often handle peak discharge rates of 3C–5C for high-power bursts, though sustained high rates can increase internal resistance and lead to thermal stress. The longevity of the battery is tracked through its cycle life, with many 2025 LFP (Lithium Iron Phosphate) models capable of exceeding 3000–6000 full cycles before their capacity drops below 80% of the original rating [18]. While temperature extremes can impact these metrics, cold weather slows ion transport, whereas weather accelerates chemical degradation. The integration of advanced BMS ensures these characteristics remain stable across a wide operating range of −20 °C to 60 °C.

2.11 States of Battery

LIBs are characterized by various states and metrics that define their performance, health, and usability Figure 2.7. The SOC indicates the current energy level of the battery as a percentage of its total capacity, serving as a vital measure for energy management. The State of Health (SOH) reflects the battery's overall condition compared to its initial state, helping assess its long-term viability. The state of power (SOP) quantifies the battery's ability to deliver power at a specific moment,

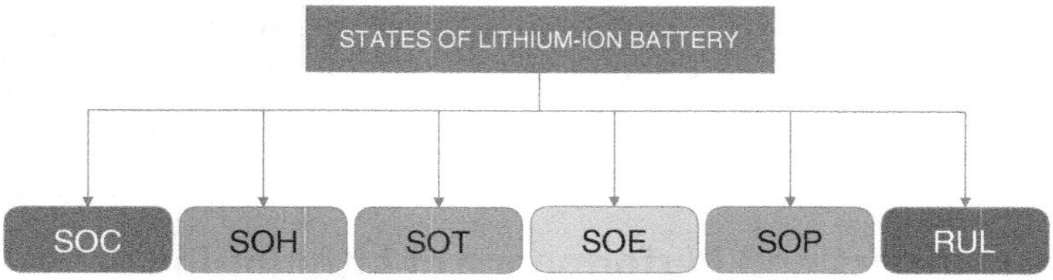

Figure 2.7 States of lithium-ion battery.

influenced by factors like temperature and internal resistance. Similarly, the SOE represents the usable energy remaining in the battery, considering efficiency and operational constraints. The depth of discharge (DOD) measures the proportion of energy discharged relative to the battery's full capacity, influencing its lifespan. The state of temperature (SOT) monitors the battery's thermal state, critical for ensuring safety and optimal performance. Lastly, the remaining useful life (RUL) estimates the duration or number of cycles the battery can operate effectively before reaching end-of-life criteria. Together, these states provide a comprehensive framework for monitoring, managing, and optimizing LIBs in diverse applications [46].

2.11.1 State of Charge

The most familiar is the SoC, which functions like a fuel gauge, it represents the current available capacity as a percentage of the battery's maximum capacity (e.g., 50% SoC means half of the available tank is full). However, this is often confused with the SoE, which is a more dynamic measure of the actual work the battery can perform [47]. While SoC tracks ampere-hours (charge), SoE tracks watt-hours (energy), accounting for the fact that battery voltage drops as it depletes. This means that 50% SoC does not always equate to exactly 50% of the remaining driving range, making SoE a more accurate predictor for EV mileage [48].

Beyond immediate energy levels, the SOH provides a long-term medical report of the battery. It compares the battery's current maximum capacity to its original capacity when brand new; for instance, an older EV might show 100% SoC (fully charged) but only 80% SOH, meaning its full state only holds 80% of the energy it did on day one. Complementing this is the SOP, which estimates the maximum burst of current the battery can safely provide at any given moment critical for tasks like rapid acceleration or regenerative braking.

2.11.2 State of Power

The SOP is not a fixed number; it is a dynamic value calculated in real-time by the BMS based on several strict operational constraints. These include voltage limits (to prevent over-voltage during charging or under-voltage during discharge), current limits (to avoid overheating), and the battery's current internal resistance.

Because internal resistance increases significantly at low temperatures or as a battery age (low SOH), the SOP will often drop in winter or toward the end of a battery's life, even if the SOC is still high. Essentially, SOP acts as a safeguard that ensures the vehicle's demands never exceed the battery's physical and chemical capabilities at any given moment [48].

2.11.3 State of Health

The SOH of a lithium-ion battery is a critical figure of merit that quantifies the battery's current condition and its remaining useful life compared to its ideal, brand-new state. Typically expressed as a percentage, a battery with an SOH of 100% matches all of its original manufacturer specifications, while a lower percentage indicates that the battery has undergone degradation. Unlike the SOC, which changes daily as you use and recharge the battery, SOH is a long-term metric that tracks permanent changes in the battery's chemistry. It is primarily measured through two indicators: capacity fade, which is the reduction in the total energy the battery can hold, and power fade, which is the increase in internal resistance that limits how quickly the battery can deliver that energy [48].

In practical applications like EVs, a battery is often considered to have reached its end of life when its SOH drops to 70% or 80%, as the reduced range and power can no longer meet the vehicle's performance requirements. The BMS calculates SOH using complex algorithms that analyze historical data, such as the number of charge cycles, temperature exposure, and voltage behavior over time. Because SOH is not a directly measurable physical quantity like voltage, different manufacturers may use slightly different formulas to estimate it, often combining capacity measurements with internal resistance checks to provide a comprehensive health score.

2.11.4 State of Temperature

The SOT often also referred to as the state of thermal condition has become a vital metric for high-performance lithium-ion systems, representing the estimated thermal status and distribution within a battery pack. Unlike simple temperature readings from external sensors, SOT involves sophisticated modeling by the BMS to estimate the core temperature of the cells, which can be significantly higher than the surface temperature during rapid charging or high-load discharging [49]. Maintaining the SOT within the Goldilocks zone (typically 20–40 °C) is essential because the battery's chemical kinetics are highly sensitive to thermal fluctuations; cold temperatures increase internal resistance and risk lithium plating, while excessive heat accelerates electrolyte decomposition and can trigger thermal runaway. By accurately predicting SOT, the BMS can proactively activate cooling or heating systems, or throttle the SOP to prevent the battery from reaching dangerous thermal limits, thereby ensuring both functional safety and a longer lifespan.

2.11.5 State of Energy

The SOE of a lithium-ion battery is a dynamic metric that represents the remaining usable energy in the battery, typically expressed as a percentage of its total potential energy. While the more common SOC acts like a simple “fuel gauge” by tracking the amount of electric charge (ampere-hours) remaining, the SOE is a more sophisticated measure that accounts for the voltage at which that charge is delivered [50]. Because the voltage of a lithium-ion cell naturally declines as it discharges, the last 10% of charge contains less total energy (watt-hours) than the first 10%. By integrating both current and voltage over time, the SOE provides a highly accurate reflection of the battery’s ability to perform actual work, making it the superior metric for predicting the remaining driving range in EVs or the true backup time for energy storage systems. In a BMS, SOE is calculated by comparing the remaining available energy to the maximum available energy the battery can hold in its current state of health. This calculation is vital because it accounts for real-time operating conditions such as temperature and discharge rates, which can cause internal resistance to rise and usable energy to drop even if the “charge” level seems high. In 2025, advanced BMS algorithms use energy counting methods (the integration of power,

$(P = V \times I)$ rather than just “coulomb counting” to ensure that users are not caught off guard by a sudden drop in power as the battery nears empty.

2.11.6 Remaining Useful Life

Remaining useful life (RUL) is the estimated time, distance, or number of charge–discharge cycles a battery can undergo before its performance falls below an acceptable functional threshold the end of life (EOL). In most automotive and industrial standards, EOL is reached when the battery’s capacity drops to 70% or 80% of its original value [51].

2.12 Solved Examples on Calculating Battery Capacity, Estimation of Battery Lifetime

While SOH tells you the battery’s condition right now, RUL is a predictive forecast. It uses your current driving habits, charging frequency, and environmental temperatures to tell you, for example, your battery is currently healthy, but at your current usage rate, it will need replacement in 3.5 years.

Calculating battery capacity (Q) is the total amount of electric charge a battery can store, typically measured in Ampere-hours (Ah). The formula:

$$Q = I \times t$$

where I is current (Amps) and t is time (Hours).

Problem 1: A technician discharges a laptop battery at a constant current of 500 mA (0.5 A). It takes exactly 8 hours and 15 minutes for the battery to reach its cut-off voltage. What is the measured capacity?

Solution: Convert time to hours: 15 minutes = 0.25 hours. Total time $t = 8.25$ hours. Apply the formula: $Q = 0.5$ A × 8.25 hours Result: $Q = 4.125$ Ah (or 4125 mAh).

Estimating Battery Lifetime (Runtime): Runtime refers to how long a specific device can run on a single charge. The formula:

$$\text{Runtime (h)} = \frac{\text{Battery capacity (Wh)}}{\text{Device power consumption (W)}} \times \text{Efficiency}$$

Problem 2: You have a 12 V e-bike battery with a 500 Wh capacity. The motor consumes an average of 250 W during a ride. Assuming a 90% system efficiency (due to heat loss in the motor and controller), how long will the battery last?

Solution: Calculate ideal runtime: 500 Wh ÷ 250 W = 2 hours. Account for efficiency: 2 hours × 0.90 = 1.8 hours.

Result: The battery will last 1 hour and 48 minutes.

Estimating total battery life (Cycles): Total life estimation predicts how many cycles a battery can handle before it reaches its EOL threshold.

Problem 3: A lithium-ion battery is rated for 2000 cycles at a 100% DoD. If a user only discharges the battery to 50% (0.5 DoD) every day before recharging, and research shows these triples the cycle life, how many years will the battery last?

Step-by-Step Solution: Calculate new cycle life: 2000 cycles × 3 = 6000 cycles. Calculate years: 6000 cycles ÷ 365 days/year ≈ 16.4 years. Result: By shallow discharging the battery, the estimated life is approximately 16.4 years.

References

1 M. S. Whittingham, "History, evolution, and future status of energy storage," in *Proceedings of the IEEE, Institute of Electrical and Electronics Engineers Inc.*, pp. 1518–1534, 2012. https://doi.org/10.1109/JPROC.2012.2190170.

2 S. G. Leonardi, M. Samperi, L. Frusteri, V. Antonucci, and C. D'Urso, "A review of sodium-metal chloride batteries: Materials and cell design," *Batteries*, vol. 9, no. 11, p. 524, 2023. https://doi.org/10.3390/batteries9110524.

3 V. Anilan, A. Vij, and H. Barrie, "Why some countries lead and others lag? The political economy of promoting battery electric vehicles," 2025. https://doi.org/10.2139/ssrn.5408390.

4 V. T. D. Sang, Q. H. Duong, L. Zhou, and C. F. A. Arranz, "Electric vehicle battery technologies and capacity prediction: A comprehensive literature review of trends and influencing factors," *Batteries*, vol. 10, no. 12, p. 451, 2024. https://doi.org/10.3390/batteries10120451.

5 E. B. Agyekum, J. D. Ampah, T. Wilberforce, S. Afrane, and C. Nutakor, "Research progress, trends, and current state of development on PEMFC-new insights from a bibliometric analysis and characteristics of two decades of research output," *Membranes (Basel).*, vol. 12, no. 11, p. 1103, 2022. https://doi.org/10.3390/membranes12111103.

6 A. ur Rehman and M. Shakeel Sadiq Jajja, "Strategic adaptation in the electric vehicle supply chain: navigating transformative trends in the automobile industry," *J. Enterp. Inf. Manag.*, vol. 38, no. 3, pp. 745–767, 2025. https://doi.org/10.1108/JEIM-10-2024-0554.

7 J. Sun, S. Chen, K. Shen, and Y. Zheng, "Liquid cooling system optimization for a cell-to-pack battery module under fast charging," *Int. J. Energy Res.*, vol. 46, no. 9, pp. 12241–12253, 2022. https://doi.org/10.1002/er.7990.

8 M. Sethurajan et al., "Recent advances on hydrometallurgical recovery of critical and precious elements from end of life electronic wastes—a review," *Crit. Rev. Environ. Sci. Technol.*, vol. 49, no. 3, pp. 212–275, 2019. https://doi.org/10.1080/10643389.2018.1540760.

9 S. S. Madani et al., "Advances in battery modeling and management systems: A comprehensive review of techniques, challenges, and future perspectives," *Batteries*, vol. 11, no. 11, p. 426, 2025. https://doi.org/10.3390/batteries11110426.

10 G. Dotson, "Congress's fifty year mission to transition motor vehicles: A brief history of federal electric vehicle policy in the United States," *SSRN Electronic Journal*, 2025. https://doi.org/10.2139/ssrn.5256344.

11 C. A. MacCarley, "Electric vehicles," In: *Non-Petroleum Automotive Transportation*, Cham: Springer Nature Switzerland, pp. 391–572, 2025. https://doi.org/10.1007/978-3-031-85028-8_14.

12 E. J. Van Der Hout, "Striking gold in the valley of death: identifying key drivers of venture capital investment in emergent sustainable technologies graduation committee."

13 U. S. Meda, L. Lal, S. M, and P. Garg, "Solid Electrolyte Interphase (SEI), a boon or a bane for lithium batteries: A review on the recent advances," *J. Energy Storage*, vol. 47, p. 103564, 2022, https://doi.org/10.1016/j.est.2021.103564.

14 S. Kiemel, et al., "Assessing the application-specific substitutability of lithium-ion battery cathode chemistries based on material criticality, performance, and price," *Resources*, vol. 10, no. 9, p. 87, 2021. https://doi.org/10.3390/resources10090087.

15 M. Apostolopoulou, D. Vernardou, and S. Passerini, "Optimization strategies of hybrid lithium titanate oxide/carbon anodes for lithium-ion batteries," *Nanomaterials*, vol. 14, no. 22, p. 1799, 2024. https://doi.org/10.3390/nano14221799.

16 A. Machín and F. Márquez, "The next frontier in energy storage: A game-changing guide to advances in solid-state battery cathodes," *Batteries*, vol. 10, no. 1, p. 13, 2023. https://doi.org/10.3390/batteries10010013.

17 X. Chen, W. Shen, T. T. Vo, Z. Cao, and A. Kapoor, "An overview of lithium-ion batteries for electric vehicles," in *2012 10th International Power & Energy Conference (IPEC)*, IEEE, pp. 230–235, 2012. https://doi.org/10.1109/ASSCC.2012.6523269.

18 I. T. Adebanjo et al., "A comprehensive review of lithium-ion battery components degradation and operational considerations: A safety perspective," *Energy Adv.*, vol. 4, no. 7, pp. 820–877, 2025. https://doi.org/10.1039/D5YA00065C.

19 K. Xu, "Electrolytes and interphases in Li-ion batteries and beyond," *Chem. Rev.*, vol. 114, no. 23, pp. 11503–11618, 2014, https://doi.org/10.1021/cr500003w.

20 W. Zhang, Z. Liang, X. Yin, and G. Ling, "Avoiding thermal runaway propagation of lithium-ion battery modules by using hybrid phase change material and liquid cooling," *Appl. Therm. Eng.*, vol. 184, p. 116380, 2021, https://doi.org/10.1016/j.applthermaleng.2020.116380.

21 T. N. V. Krishna, S. V. S. V. P. D. Kumar, S. Srinivasa Rao, and L. Chang, "Powering the future: Advanced battery management systems (BMS) for electric vehicles," *Energies (Basel).*, vol. 17, no. 14, p. 3360, 2024. https://doi.org/10.3390/en17143360.

22 F. Yeganehdoust, A. K. Madikere Raghunatha Reddy, and K. Zaghib, "Cell architecture design for fast-charging lithium-ion batteries in electric vehicles," *Batteries*, vol. 11, no. 1, p. 20, 2025. https://doi.org/10.3390/batteries11010020.

23 J. Cabana, L. Monconduit, D. Larcher, and M. R. Palacín, "Beyond intercalation-based Li-ion batteries: The state of the art and challenges of electrode materials reacting through conversion reactions," *Advanced Materials*, vol. 22, no. 35, 2010. https://doi.org/10.1002/adma.201000717.

24 I. Langmuir, "The Interaction of electron and positive ion space charges in cathode sheaths," *Phys. Rev.*, vol. 33, no. 6, pp. 954–989, 1929. https://doi.org/10.1103/PhysRev.33.954.

25 H. Du et al., "Side Reactions/Changes in Lithium-Ion Batteries: Mechanisms and Strategies for Creating Safer and Better Batteries," *Adv. Mater.*, vol. 36, no. 29, 2024. https://doi.org/10.1002/adma.202401482.

26 F. Cheng, J. Liang, Z. Tao, and J. Chen, "Functional materials for rechargeable batteries," *Adv. Mater.*, vol. 23, no. 15, pp. 1695–1715, 2011. https://doi.org/10.1002/adma.201003587.

27 X. Zhang et al., "Recent advances in functionalized separator for dendrites-free and stable lithium metal batteries," *EcoEnergy*, vol. 2, no. 4, pp. 549–598, 2024. https://doi.org/10.1002/ece2.58.

28 M. Z. Bazant, "Theory of chemical kinetics and charge transfer based on nonequilibrium thermodynamics," *Acc. Chem. Res.*, vol. 46, no. 5, pp. 1144–1160, 2013. https://doi.org/10.1021/ar300145c.

29 S. Srinivasan, "Electrode/electrolyte interfaces: Structure and kinetics of charge transfer," in *Fuel Cells*, Boston, MA: Springer US, pp. 27–92, 2006. https://doi.org/10.1007/0-387-35402-6_2.

30 P. Bai and M. Z. Bazant, "Charge transfer kinetics at the solid–solid interface in porous electrodes," *Nat. Commun.*, vol. 5, no. 1, p. 3585, 2014. https://doi.org/10.1038/ncomms4585.

31 J. Mangaiyarkkarasi and S. R. J., "Nanostructural innovations," pp. 51–74, 2024. https://doi.org/10.4018/979-8-3693-5320-2.ch003.

32 G.-J. Chen, C.-L. Liu, Y.-H. Liu, and J.-J. Wang, "Implementation of constant temperature–constant voltage charging method with energy loss minimization for lithium-ion batteries," *Electronics (Basel).*, vol. 13, no. 3, p. 645, 2024. https://doi.org/10.3390/electronics13030645.

33 A. B. Khan and W. Choi, "Optimal charge pattern for the high-performance multistage constant current charge method for the Li-ion batteries," *IEEE Trans. Energy Convers.*, vol. 33, no. 3, pp. 1132–1140, 2018. https://doi.org/10.1109/TEC.2018.2801381.

34 A. Eddahech, O. Briat, and J.-M. Vinassa, "Determination of lithium-ion battery state-of-health based on constant-voltage charge phase," *J. Power Sources*, vol. 258, pp. 218–227, 2014. https://doi.org/10.1016/j.jpowsour.2014.02.020.

35 M. D. Levi et al., "Solid-state electrochemical kinetics of Li-ion intercalation into Li1 − x CoO2: Simultaneous application of electroanalytical techniques SSCV, PITT, and EIS," *J. Electrochem. Soc.*, vol. 146, no. 4, pp. 1279–1289, 1999. https://doi.org/10.1149/1.1391759.

36 S. Zhao, F. Wu, L. Yang, L. Gao, and A. F. Burke, "A measurement method for determination of dc internal resistance of batteries and supercapacitors," *Electrochem. commun.*, vol. 12, no. 2, pp. 242–245, 2010. https://doi.org/10.1016/j.elecom.2009.12.004.

37 M. Shen and Q. Gao, "A review on battery management system from the modeling efforts to its multiapplication and integration," *Int. J. Energy Res.*, vol. 43, no. 10, pp. 5042–5075, 2019. https://doi.org/10.1002/er.4433.

38 Z. B. Omariba, L. Zhang, and D. Sun, "Review of battery cell balancing methodologies for optimizing battery pack performance in electric vehicles," *IEEE Access*, 2019. https://doi.org/10.1109/ACCESS.2019.2940090.

39 P. K, A. V Amaragatti, Y. G. L, P. Manohar, S. V Kulkarni, and K. S. Ramanujan, "Real time battery monitoring and protection system with can bus communication and data logging," in *2024 IEEE 4th International Conference on Sustainable Energy and Future Electric Transportation (SEFET)*, IEEE, pp. 1–6, 2024. https://doi.org/10.1109/SEFET61574.2024.10718082.

40 H. Gabbar, A. Othman, and M. Abdussami, "Review of battery management systems (BMS) development and industrial standards," *Technologies (Basel).*, vol. 9, no. 2, p. 28, 2021. https://doi.org/10.3390/technologies9020028.

41 N. N. Pham, J. Leuchter, K. L. Pham, and Q. H. Dong, "Battery management system for unmanned electric vehicles with CAN BUS and internet of things," *Vehicles*, vol. 4, no. 3, pp. 639–662, 2022. https://doi.org/10.3390/vehicles4030037.

42 L. A. Perişoará, I. B. Bacîş, and D. I. Săcăleanu, "Master-slave BMS architecture with CAN-bus for inter-cell communication," in *2024 IEEE 30th International Symposium for Design and Technology in Electronic Packaging (SIITME)*, IEEE, pp. 227–232, 2024. https://doi.org/10.1109/SIITME63973.2024.10814795.

43 Z. Liu, C. Tan, and F. Leng, "A reliability-based design concept for lithium-ion battery pack in electric vehicles," *Reliab. Eng. Syst. Saf.*, vol. 134, pp. 169–177, 2015. https://doi.org/10.1016/j.ress.2014.10.010.

44 M. Wang, S. Wu, Y. Chen, and W. Luan, "The snowball effect in electrochemical degradation and safety evolution of lithium-ion batteries during long-term cycling," *Appl. Energy*, vol. 378, p. 124909, 2025. https://doi.org/10.1016/j.apenergy.2024.124909.

45 B. A. Johnson and R. E. White, "Characterization of commercially available lithium-ion batteries," *J. Power Sources*, vol. 70, no. 1, pp. 48–54, 1998. https://doi.org/10.1016/S0378-7753(97)02659-1.

46 L. Zhou et al., "State estimation models of lithium-ion batteries for battery management system: Status, challenges, and future trends," *Batteries*, vol. 9, no. 2, p. 131, 2023. https://doi.org/10.3390/batteries9020131.

47 W. Zhou, Y. Zheng, Z. Pan, and Q. Lu, "Review on the battery model and SOC estimation method," *Processes*, vol. 9, no. 9, p. 1685, 2021. https://doi.org/10.3390/pr9091685.

48 P. Reshma and V. J. Manohar, "Collaborative evaluation of SoC, SoP and SoH of lithium-ion battery in an electric bus through improved remora optimization algorithm and dual adaptive Kalman filtering algorithm," *J. Energy Storage*, vol. 68, p. 107573, 2023. https://doi.org/10.1016/j.est.2023.107573.

49 Y. Xie et al., "Coestimation of SOC and three-dimensional SOT for lithium-ion batteries based on distributed spatial–temporal online correction," *IEEE Transactions on Industrial Electronics*, vol. 70, no. 6, pp. 5937–5948, 2023. https://doi.org/10.1109/TIE.2022.3199905.

50 H. He, Y. Zhang, R. Xiong, and C. Wang, "A novel Gaussian model based battery state estimation approach: State-of-Energy," *Appl. Energy*, vol. 151, pp. 41–48, 2015. https://doi.org/10.1016/j.apenergy.2015.04.062.

51 S. A. Hasib et al., "A comprehensive review of available battery datasets, RUL prediction approaches, and advanced battery management," *IEEE Access*, vol. 9, pp. 86166–86193, 2021. https://doi.org/10.1109/ACCESS.2021.3089032.

3

Alternative Energy-storage Devices: Supercapacitors and Fuel Cell

3.1 Introduction

The development of electrochemistry and electrical storage devices has been multifaceted and branching beyond batteries such as supercapacitors and fuel cells. Each of the respective technologies has been capitalized on different methodologies of electrochemistry, resulting in different capabilities, strengths, and weaknesses. All these methods have gone through and will continue to have advancements through research, which will be discussed within this chapter. To fully appreciate the capabilities and drawbacks of alternative storage devices, an understanding of the chemical and mechanical properties used to make the devices must be appreciated.

Reviewing other means of energy storage, primarily mechanical and chemical systems have been used to store some energy as is categorized in Figure 3.1. Mechanical systems, such as gravity-based solid-mass storage, hydroelectric storage, compressed-air energy storage, and flywheels, typically utilize potential energy through gravity or rotational means prior to running a generator. Some thermal systems, such as latent-heat thermal energy-storage systems (e.g. steam accumulator) and cryogenic energy-storage systems, may utilize turbines to convert energy stored in water or ice into electrical energy. Carnot batteries store electricity as heat, utilizing thermodynamic cycles such as the Rankine cycle or Brayton cycle. Although the variation of mechanical energy storage is quite vast, the majority are utilized only for specialty applications due to low efficiencies ranging from 25% to 85% of conversion and recovery. Many systems also suffer from low availability, ease of electrical recoverability, and cost. The most prominent mechanical method is hydroelectric due to large availability of large bodies of water with an efficiency of around 90%. Otherwise, flywheels are widely used within automotive with assistance to a electric motor/generator system.

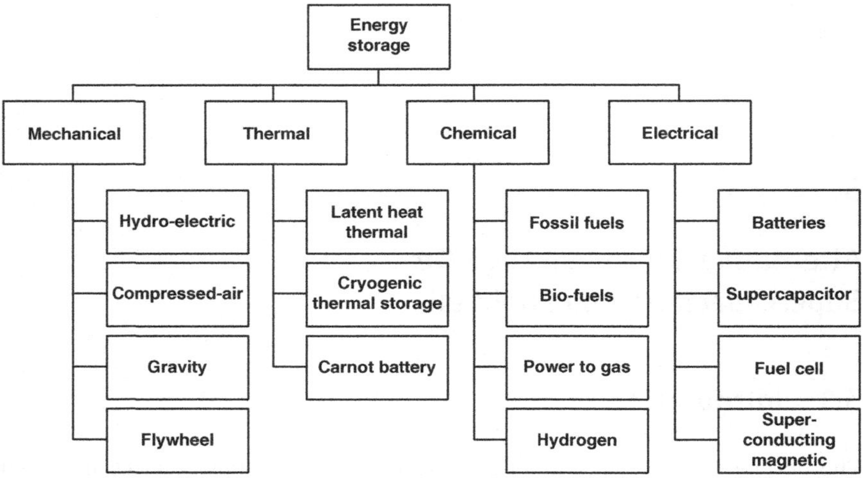

Figure 3.1 Energy-storage techniques.

Apart from mechanical methods of energy storage, chemical energy-storage systems have been the most prevalent for portable energy storage. This is inclusive of utilizing fossil fuels for energy generation, specifically for electricity, it would be in the form of generators or through an alternator system. Power-to-gas or power-to-liquid systems, which allow electricity to store energy through electrolysis, allow for a reversible storage of energy but will be discussed within the fuel cell section. Briefly, superconducting magnetic energy-storage (SMES) systems are a direct electrical energy-storage method but are unpopular due to the strict needs of the device. As the name implies, SMES store energy in a magnetic field of a coil that has been cryogenically cooled to below the coil's superconducting critical temperature. This system is typically unsuitable for most applications due to the constant upkeep and initial cost of implementing an SMES.

3.2 Supercapacitors

Supercapacitors, also known as ultracapacitors, are an electrical energy-storage system utilized to temporarily store tremendous amounts of energy for rapid charge and discharge. Supercapacitors can either function utilizing Faradaic or non-Faradaic processes resulting in three major categories; Electric double-layer capacitors (EDLCs), which store charge electrostatically, pseudocapacitor utilize faradaic/redox mechanisms for charge storage, and hybrids apply electrostatic-electrochemical with asymmetric electrode materials categorized in Figure 3.2. Fuel cell type performance characteristics as well usage considerations are compared in Table 3.1.

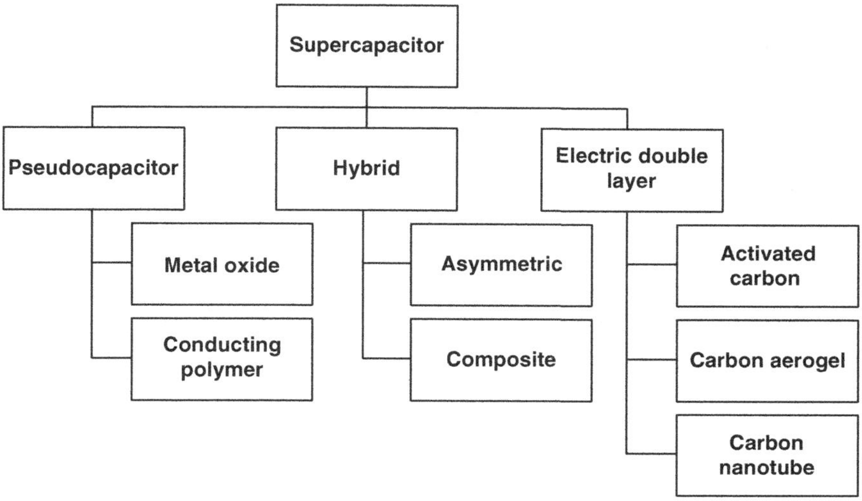

Figure 3.2 Classification of supercapacitors.

$$Q = \mathrm{CV} = \frac{Ae}{d}V \tag{3.1}$$

Since storage of charge is a surface property for supercapacitors, researchers have explored a wide variety of materials to increase the surface area, such as nanomaterials, polymers, and carbon to develop nanopore increasing surface area. Q is the energy, A is the area of plates, e is the permittivity of the dielectric, d is the distance between plates. Figure 3.3 shows a sample of a supercapacitor bank as a physical example.

3.2.1 Classical Modeling

Supercapacitors can be generically modeled using lumped parameters. In this case, the basic model can be represented by a series resistor and an RC pair show in Figure 3.4. The series resistance R_s can be calculated by the change in voltage and change in current ΔV and ΔI, respectively. R_p can be calculated with Equation (3.4).

$$R_s = \frac{\Delta V}{\Delta I} \tag{3.2}$$

$$V_2(t) = V_i\left(e^{-\frac{t}{R/C}}\right) \tag{3.3}$$

$$R_p = -\frac{t}{\ln\left(\frac{V_2}{V_1}\right)C} \tag{3.4}$$

$$C = \frac{2\int_{t_1}^{t_2} Vi\, dt}{\left(V_1^2 - V_2^2\right)} \tag{3.5}$$

Table 3.1 Supercapacitor comparison.

	Electric double layer	Pseudo	Hybrid
Anodes	Nanostructure carbon	Electrically conducting polymers Transitional metal oxides	Nanocomposites
Specific energy (Wh/kg)	4.9	9	37
Specific power (W/kg)	206,000	10,000	14,000
Efficiency	95%	95%	90%
Benefits	Simple behavior Fast kinetics Long life durability	Larger capacitance Higher energy density	Larger capacitance Selectable properties
Drawbacks	Low chemical capacitance Capacitance is tied to electrode size	Chemical degradation Slower kinetics	Chemical degradation Complex expensive
Usage	Transportation Backup power Grid stability	Transportation Portable electronics Industrial	Energy harvesting Automation

Figure 3.3 Supercapacitor bank.

where V_1 is the initial voltage, V_2 is the final voltage, t is the time, C is the capacitance, usable energy is represented by E_{usable}, ΔV is the change in voltage, and ΔI is the change in current. Equation (3.6) can be used to calculate the usable energy in both electrolytic and dielectric capacitors, with consideration for the voltage drop Equation (3.7).

$$E_{stored} = \frac{1}{2}C \times \left(V_{capacitor}^2\right) \tag{3.6}$$

$$E_{usable} = \frac{1}{2}C \times \left(V_1^2 - V_2^2\right) \tag{3.7}$$

3.2.1.1 Supercapacitor Types

3.2.1.1.1 Electrical-double layer

EDLC also known as electrochemical capacitors, store charge at both interfaces between the anode and electrolyte and cathode and electrolyte resulting in the double-layered electrochemical charge formation as shown in Figure 3.5. While charging the electrostatic force

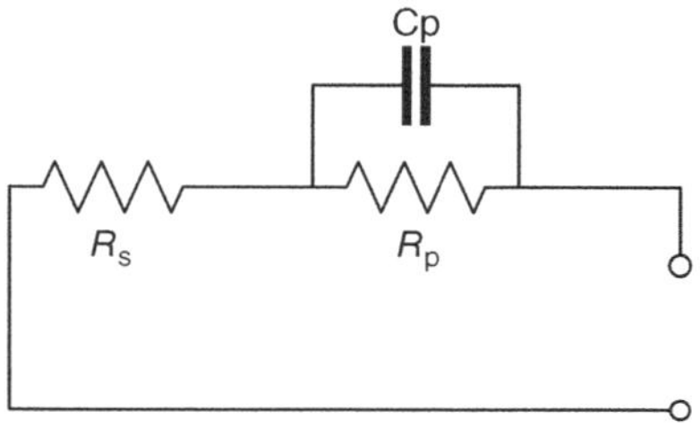

Figure 3.4 1RC classical equivalent model.

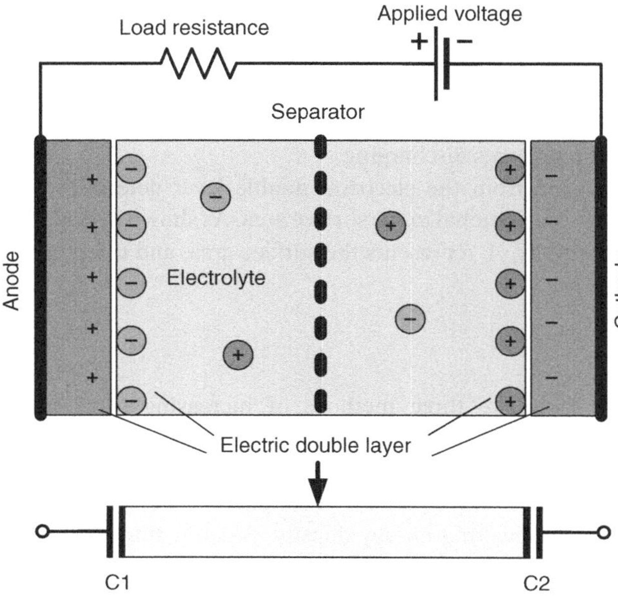

Figure 3.5 Construction of a charging EDLC.

of attraction pulls the electric charges towards the charged plate with opposing charge. The electrolyte ions are then accumulated at the interface between the electrolyte and electrodes resulting in the formation of the electric double layer (EDL).

The charge is stored through the formation of the EDL and not through a Faradaic or charge transfer process. This results in a rechargeable component with an extremely high cyclic count (<10,000 cycles), as the cycling does not involve a chemical reaction process [1]. This is possible because, during charging, an electronic charge layer forms on the surface of the electrode. As the electrolyte ions are free floating within the electrolyte, ions of the opposite polarity are attracted to the charged layer while remaining within the electrolyte.

Within the electrolyte itself, the distribution of the ions can be modeled as two regions: the inner Helmholtz and the outer Helmholtz layer. The inner Helmholtz layer is at the electrolyte//electrode interface. This layer contains densely packed ions due to electrostatic attraction forming a region of charged particles. Within this layer, the potential drops linearly as it approaches the second outer Helmholtz layer further into the electrolyte. Finally, from the outer Helmholtz layer, the potential continues to decay exponentially until the diffuse layer, where the ions are still free floating [2].

$$E_{\mathrm{p}} + E_{\mathrm{n}} + (I^- + I^+) \Leftrightarrow E_{\mathrm{p}}^+//I^- + E_{\mathrm{n}}^-//I^+ \tag{3.8}$$

The reversible process of the EDLC can be shown as Equation (3.8) with E_{p} and E_{n} as the positive and negative electrodes, respectively. The cation and anion are represented as I^- and I^+, respectively, while // represents the double-layer interface. The process shown on the right represents charging, whereas the process shown on the left represents discharging.

The surface area developed from the electrical double layer determines the capacitance, as it is linearly proportional to the surface area. As shown in Equation (3.9), where ε is the permittivity, A_{s} represents the surface area, and d represents the plate-separation distance.

$$C = \frac{\varepsilon A_{\mathrm{s}}}{d} \tag{3.9}$$

Due to this property, there are three methods of increasing capacitance: increasing the permittivity, decreasing the distance between the plates or increasing the surface area. An electrolyte with high permittivity or high ionic conductivity is typically sought; however, if it is too conductive, it may reduce the potential window, ultimately lowering energy density. Broadly, the electrolytes suitable for supercapacitors are aqueous and organic electrolytes. Aqueous electrolytes offer higher ionic conductivity and therefore, higher capacitance, whereas organic electrolytes allow for a greater potential window and energy density [3].

Surface area can be selected from materials with large internal structures, such as activated carbon, which can have a surface area greater than 2000 m^2/g [4]. Current research advances in nanomaterials, such as carbon nanotubes, graphene, and other carbon structures have permitted even higher surface areas. To maximize the gains from material science for both the electrode material and electrolyte, the distance between the plates is required to be as small as possible. The distance is then calculated from the breakdown voltage of the battery (V_{bd}) and the dielectric strength of the electrolyte (E_{ds}) shown in Equation (3.10).

$$V_{bd} = E_{ds} \times d \tag{3.10}$$

To increase the durability of the supercapacitor, the distance must be selected such that the breakdown voltage is sufficiently large enough to accommodate nominal operating voltages. If the breakdown voltage is too low, the electrolyte may break down and become conductive, resulting in reduced capacity, component failure, and a short circuit [5]. The power of a supercapacitor can be calculated using Equation (3.11).

$$P = \frac{1}{4}\frac{V^2}{R_s} \tag{3.11}$$

3.2.1.1.2 Pseudo/redox

Pseudocapacitors utilize Faradaic or reduction/oxidation reactions as the method of charge storage and transfer [6]. The reversible redox reactions occur at the electrode to allow for a reusable supercapacitor and a long life cycle. Due to the operational nature of supercapacitors, the electrodes experience continuous and rapid chemical reactions. The ions do not experience any change since the redox is only at the surface of the electrodes and the only process is charge transfer shown in Figure 3.6. To allow for the continuous use of the electrodes, many materials

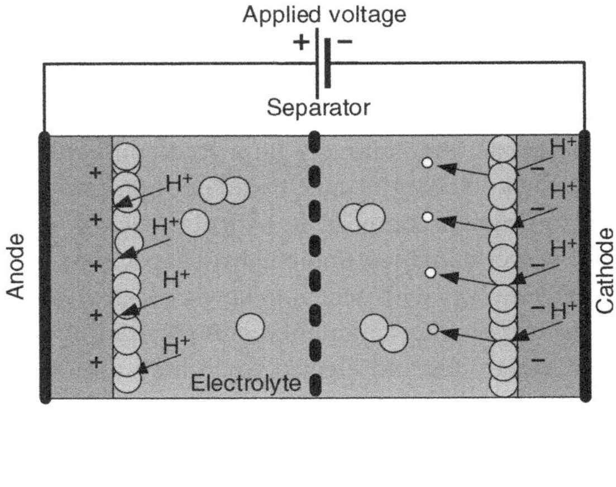

Figure 3.6 Pseudocapacitor.

have been researched for varying efficacy such as conductive polymers, metal oxides, and nanocomposites. Materials with high porosity or microstructures are preferred to allow for higher diffusion of electrolyte ions into the structures. Pseudo capacitors operate from three chemical mechanisms to allow for charge transfer absorbative, redox, and intercalation capacitance.

Adsorption capacitance encompasses the absorption and release of electrolyte ions into the microstructure material of the electrode. Reduction/oxidation is the primary mechanism of charge transfer in a pseudocapacitance operation on Faradaic charge transfer. The final mechanism is intercalation capacitance, in which electrolyte ion intercalate into the lattice of the electrode through the Van der Waal gaps. Typically, pseudocapacitors have good performance and high capacitance but have lower power density compared to electric double-layer supercapacitors.

3.2.1.1.3 Hybrid

Hybrid capacitors exhibit properties of both EDLCs and pseudocapacitors due to the use of dissimilar or composite electrode materials. The principle behind the development of hybrid capacitors is to attempt to maximize the energy density by using both existing supercapacitor principles to achieve the objective. This has allowed for higher specific capacitance and potential window compared to supercapacitors with similar electrode materials [7]. Hybrid capacitors can be categorized into three categories: composite electrode, asymmetric-electrode, and battery-type electrode.

Composite electrodes combine carbon materials with pseudocapacitive materials. Battery hybrids utilize a battery electrode along with the typical carbon electrode. Asymmetric capacitors use differing properties for operation of the electrodes, one will utilize redox materials similar to pseudocapacitors and the other electrode will behave like an EDLC as illustrated in Figure 3.7. The benefits of a hybrid capacitor are up to the selection of the property that is focused on ranging from high specific capacitance, high potential window, and high energy density depending on the construction and composite type [8].

As hybrids are the newest generation of supercapacitors and offer the largest flexibility in design and available materials, the performance of hybrid supercapacitors is largely improved compared to either EDLC or PC. This is due to the ability to mix Faradaic, non-faradaic, and electrostatic processes into a singular component, hybrid capacitor will be able to have higher energy and power densities. Some hybrid capacitors are also able to operate in higher voltage ranges with access to newer electrolytes. The main drawback of hybrid capacitors is the heightened complexity due to the mixture of more specialized electrodes and electrolyte, which also incurs higher costs compared to the ELDC.

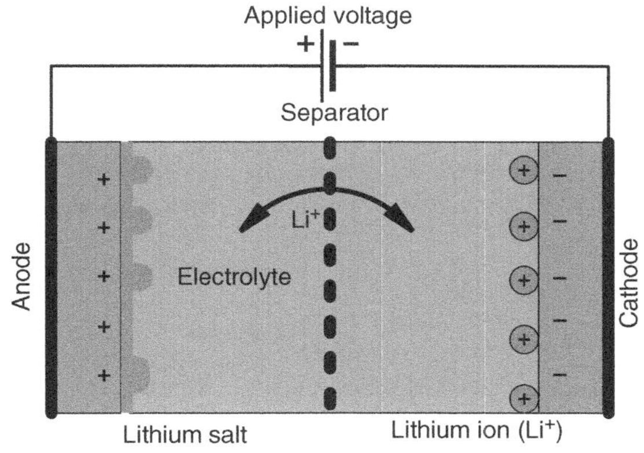

Figure 3.7 Hybrid capacitor.

Example 1: Consider a load requiring 20 W for 180 seconds. The capacitor operating voltage is 4.2 V. Solve for capacitance (C). The starting voltage is 4.2 V and minimum voltage is 3.3 V. The internal resistance is 0.036 Ω.

$$E = P \times t = 20 \times 180 = 3600\,J \tag{3.12}$$

$$C_{\text{ideal}} = 2 \times \frac{E}{V^2} = 2 \times \frac{3600}{(4.2)^2} = 408F \tag{3.13}$$

$$T_{\text{actual,ideal}} = \frac{1}{2} \times p \times c \times \left(V_{\text{start}}^2 - V_{\text{min}}^2\right) = 68\text{ s} \tag{3.14}$$

$$I = \frac{20}{4.2} = 4.76\text{ A} \tag{3.15}$$

$$V_{\text{drop}} = I \times R_{\text{internal}} = 0.17\text{ V} \tag{3.16}$$

$$V_{\text{start}} = V_{\text{initial}} - V_{\text{drop}} = 4.03\text{ V} \tag{3.17}$$

$$C = \frac{2 * (t * P)}{(V_{\text{start}}^2 - V_{\text{min}}^2)} = 1340\ F \tag{3.18}$$

3.3 Fuel Cells

A fuel cell is an electrochemical cell, which can continuously convert chemical energy to electrical energy through oxidation utilizing fuels with high concentration of Hydrogen such as alcohols, pure hydrogen and hydrocarbons.

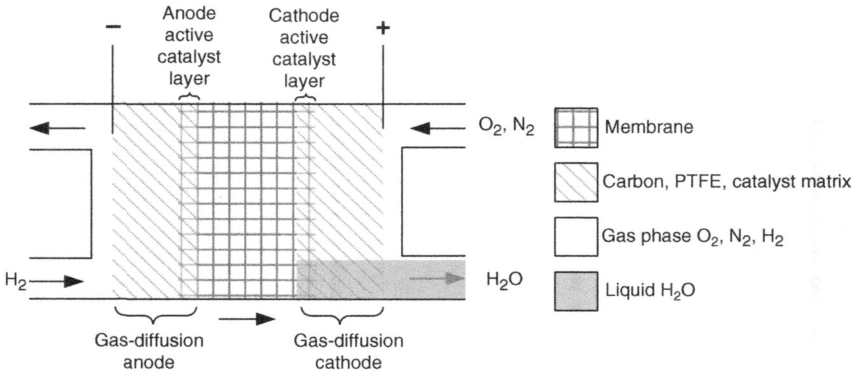

Figure 3.8 Cross section of a fuel cell.

Utilizing ideal components such as pure Hydrogen and Oxygen, a basic fuel cell will keep the two reactants separate with a separation membrane filled with an electrolyte as the only medium of contact shown in Figure 3.8. This allows for higher control of the energy production as simple oxidation of the Hydrogen at the Anode and reduction of the Oxygen at the cathode. Hydrogen permeates the gas diffusion anode and at the anode active catalyst later, the hydrogen is oxidized with assistance from a catalyzer to produce electrons (e^-) and Hydrogen ions (H^+). The electrons will be accumulated at the anode current collector to flow through the load in an external circuit then the cathode current collector generates the electrical power. The protons generated from the oxidation of hydrogen will travel internally through the electrolyte to the cathode, where the electrons from the external circuit will accumulate. At the cathode active catalyst layer, Oxygen will react with the Hydrogen ions from the electrolyte and the electrons from the cathode through a reduction reaction to produce water and heat. The overall reaction can be shown as an oxidation of hydrogen and the reduction of oxygen, with electrical energy generated and water produced as a safe byproduct of the reaction shown in Equation (3.19).

$$2H_2 + 2O^{2-} \rightarrow 2H_2O + 4e^- \tag{3.19}$$

As the electrochemical process is highly controlled redox with the reactants separated, this allows for much higher efficiency compared to other fuel-based energy-storage system such as an internal combustion engine (ICE). ICE ay have a peak efficiency of 10–40%; due to the nature of uncontrolled combustion, most of the energy is lost in heat, light, incomplete combustion, and an additional mechanical process is required to convert the energy released during combustion into usable energy. Comparatively, fuel cells are an electrochemical process allowing for direct conversion into usable electricity, the efficiency ranges from 40% to 80%,

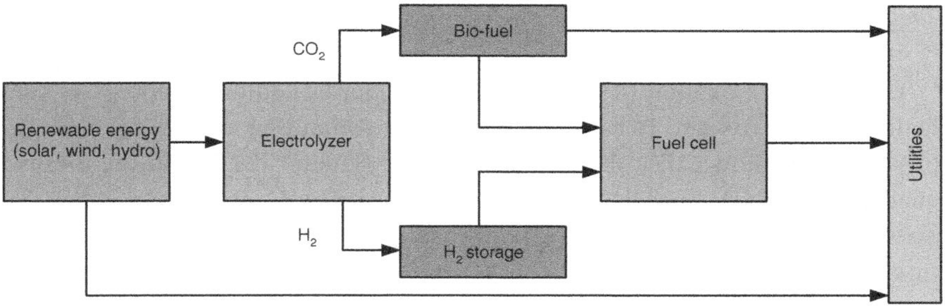

Figure 3.9 Renewable power fuel chain.

depending on the type of fuel cell and whether the efficiency also includes the fuel production.

As shown in Figure 3.9, a renewable power system utilizes electrolyzers as the central connecting piece from renewable energies such as solar, hydro, and wind. At the electrolyzers, the main product would be hydrogen, allowing for cleaner energy production and energy storage from the fuel cell. Byproducts from the electrolyzers would be CO_2 and oxygen, oxygen would be usable as the oxidant in the fuel cell, while the CO_2, with more processing, would be usable with a biofuel system that could still feed the fuel cell, depending on the fuel cell type.

3.3.1 Hydrogen/fuels

Since fuel cells keep the reactants separated from each other during idle, and use the electrochemical process of reduction and oxidation, fuels must be fed to the fuel cell for continuous operation. Depending on the type of fuel cell, some can accept wider range of fuels ranging from biofuels like ethanol and methanol, ammonia, to fossil fuels such as gasoline or diesel. Although such biofuels and fossil fuels have high energy density ranging from 4.5 to 9.7 kWh/L, hydrogen as the smallest element is highly compressible allowing for a high-power density of 32.94 kW/kg compared to Diesels 12.77 kW/kg. Another benefit of utilizing pure hydrogen is with the oxidation, Oxygen, the only exhaust from the system would be water as shown in Equation (3.20).

$$2H_2 + 2O^{2-} \rightarrow 2H_2O + 4e^- \tag{3.20}$$

To generate hydrogen and oxygen, the main fuels for most fuel cells, both gases can be generated from the process of water electrolysis, which is the reverse reaction of hydrogen and oxygen combustion. It consists of two reactions: oxygen evolution reaction at the anode and hydrogen evolution at the cathode. The overall formula is shown in Equation (3.21).

$$H_2O + \text{energy} \rightarrow H_2 + \frac{1}{2}O_2 \tag{3.21}$$

As shown in Equation (3.21), the reaction requires additional energy to split hydrogen from oxygen. To calculate the energy demand for the reverse enthalpy of water formation (ΔH), the electrical energy demand (ΔG), and the thermal energy demand (Q) must be found, as ΔH is the sum of the electrical and thermal energy demands. This can be shown in Equation (3.22) under constant temperature and pressure.

$$\Delta H = \Delta G + Q \tag{3.22}$$

The thermal energy demand can be rewritten as $(T \times \Delta S)$, where T as the temperature and ΔS representing the entropy change. To calculate entropy change, the sum of the entropy of hydrogen (S_{H_2}) and oxygen (S_{O_2}) minus the entropy of water (S_{H_2O}) is shown in Equation (3.23).

$$\begin{aligned} \Delta S &= S_{H_2} + 0.5\ S_{O_2} - S_{H_2O} \\ \Delta S &= (0.1306) + 0.5(0.2051) - (0.0699) = 0.1631\ \text{kJ/mol*K} \end{aligned} \tag{3.23}$$

Using a temperature of 298.15 K, the total thermal energy demand is 48.7 kJ/mol. To find the electrical energy demand, Equation (3.22) can be rearranged to find ΔG using the enthalpy of water formation (ΔH), 385.8 kJ/mol.

$$\begin{aligned} \Delta G &= \Delta H - Q \\ \Delta G &= 285.8 - 48.7 = 237.2\ \text{kJ/mol} \end{aligned} \tag{3.24}$$

To calculate the minimum or reversible voltage for water electrolysis is shown in Equation (3.25), where n is the number of electrons, and F is Farad's number.

$$V^0 = \frac{\Delta G}{n\text{F}} = \frac{237.2 * 1000}{2 * 96485} = 1.23\text{V} \tag{3.25}$$

Therefore, water electrolysis can only commence when the cell voltage is larger than the minimum cell voltage of 1.23 V. Table 3.2 compiles the different operational characteristics of commonly used fuels such as gaseous and liquid hydrogen, alcohols, ammonia, methane, and fossil fuels such as gasoline and diesel.

3.4 Fuel Cell Types

Fuel cells are available in a wide variety of configurations, with differing electrolytes and electrode materials. This has made various methods of classification such as by temperature, fuels, and by electrolyte shown in Figure 3.10. Broadly, there are three ranges of temperature classification: low-temperature (20–100 °C), intermediate-temperature (100–600 °C), and high-temperature (500–1000 °C)

Table 3.2 Fuel comparison.

Fuel	Energy density (kWh/L)	Power density (kW/kg)	Storage (°C)/P (MPa)
H_2 gaseous	1.29	32.94	25/70
H_2 liquid	2.33	32.94	−253/0.1
Methanol (CH_3OH)	4.5	6.1	25/0.1
Ethanol (CH_3CH_2OH)	6.28	8	25/0.1
Ammonia	1.19	5.2	−33.34/1.5
Methane	6.94	13.89	−162/25
Gasoline	9.16	12.5	25/0.1
Diesel	9.72	12.77	25/0.1

Source: Data from [9].

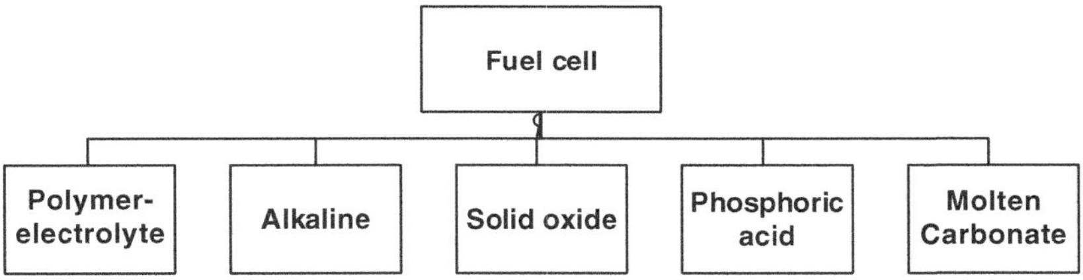

Figure 3.10 Fuel cell types.

fuel cells. Operating at higher temperatures allows for higher CO contamination, which may block pores in lower temperature fuel cells, this also allows for cheaper metals to be used for the electrodes such as nickel instead of platinum. The clear drawback is the heavy thermal control that must be exercised to maintain the higher temperatures. The fuel cell classifications based on electrolytes and fuels are typically grouped together because both are heavily dependent on each other illustrated in Figure 3.11. A summary table of performance characteristics as well as functional advantages and drawbacks are shown in Table 3.3.

3.4.1 Polymer-electrolyte Membrane Fuel Cells

Polymer-electrolyte membrane fuel cells (PEMFCs) utilize an electrolyte material made from polymers to conduct the H^+ ions. The most prominent example is Nafion, a perfluorosulfonic acid (PFSA) membrane developed by DuPont. Nafion has beneficial properties to act as a membrane such as resistance to hydrogen and oxygen crossover, high mechanical, thermal, and chemical stability as well

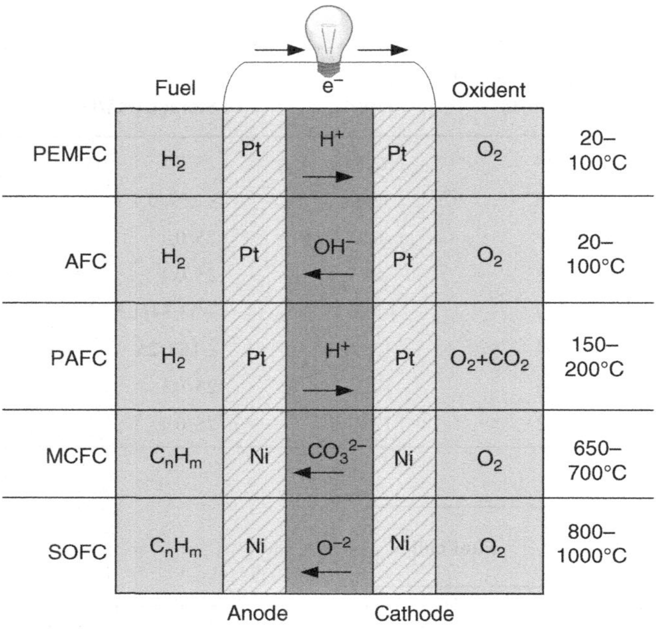

Figure 3.11 Fuel cell types, fuel, and oxidants.

as high ionic conductivity above 0.1 S/cm [10]. The material needs high levels of hydration to function nominally, resulting in a maximum operating limit for PEMFCs to be between 80 °C and 100 °C. As this is a lower temperature for a fuel cell, the hydrogen oxidation and oxygen reduction reactions need to use platinum as the electrocatalyst. As a combination of the high-performing electrocatalyst and Nafion, PEMFCs have excellent performance such as high-power density, quick startup, and fast response. With such benefits, PEMFCs are currently the most researched and applicable for transportation if not for the drawbacks. These issues range from the cost of utilizing rare metals, cooling, water management, and fuel impurity intolerance. Ongoing research is being conducted facilitate it into the transportation realm by reducing the percentage of platinum used, utilizing alloys and even nano-morphologies to make PEMFCs more cost and performance competitive.

3.4.2 Alkaline

Alkaline fuel cells (AFCs) use an electrolyte based on concentrated potassium hydroxide (KOH) and can operate at either high temperatures up to 250 °C with KOH concentrations of up to 85% or lower than 100 °C for lower concentrations of

Table 3.3 Fuel cell type comparison.

	Polymer electrolyte membrane	Alkaline	Molten	Phosphoric acid	Solid oxide
Operating temperature (°C)	40–100	80–260	600–800	170–210	600–1000
Electrolyte	KOH	Perfluorosulfonic acid	alkali metal carbonate	Phosphoric acid	YSZ
Charge carrier	H^+	OH^-	CO_3^-	H^+	O^-
Anodes	Pt	Pt	Ni alloy	Pt	Ni/YSZ
Efficiency	40–60%	45–60%	45–60%	45%	60%
Maximum power	1 MW	100 kW	3 MW	400 kW	2 MW
Benefits	Robust High power density Low temperature Fast startup fast response	Fast kinetics cheap	High efficiency Fuel flexibility Cheap catalyst	CO tolerant No water management	High efficiency Robust construction Cheap catalyst
Drawbacks	CO intolerance Water clogging Expensive catalyst Cooling needs	Sensitive to CO_2 fouling Complex Fuel cycling	Low power resistance Large internal resistance Slow	Poor reactivity Expensive catalyst Slow startup Low power density	Slow Needs thermal management
Usage	Transportation backup power	Space	Utility	Distributed	Utility auxiliary power

Source: Data from [15].

KOH [11]. As AFCs were developed earlier than other fuel cells, the electrolyte is a porous matrix such as asbestos and the ions within the electrolyte are OH^- ions. As the electrolyte is alkaline, this allows for a wider range of electrocatalysts such as nickel, silver, noble metals, and metal oxides. This flexibility allows AFCs to have much lower costs compared to other fuel cells. The largest issue with AFCs is poor fuel impurity intolerance and high sensitivity to carbon monoxide (CO) or carbon dioxide (CO_2) contamination from the oxidizer. Since AFCs utilize KOH, it readily reacts with CO or CO_2 form K_2CO_3 or other carbonate ions as a precipitates. The contamination of the electrolyte would block electrode pores, and the impurities would also lead to increased internal resistance of the fuel cell thereby reducing power and performance. To remedy these issues, solid alkaline anion exchange membranes (AEM) have been a possible solution by combining traditional AFC technology with a polymer akin to PEMFCs to take advantage of both beneficial properties. The polymer electrolyte will conduct anions such as OH^-, similar to AFCs, but since the cationic groups are fixed to the polymer structure, CO_2 contamination is reduced compared to AFCs.

3.4.3 Solid Oxide

Solid oxide fuel cells (SOFCs) operate at high temperatures of 600–1000 °C, which allows SOFCs to avoid using precious metals as electrocatalysts while still producing power in kW to MW range [12]. SOFCs typically use yttria-stabilized zirconia (YSZ) as a solid electrolyte as it can conduct oxygen ions $\left(O_2^-\right)$. The cathode used for typical SOFCs is porous lanthanum strontium manganite oxide and the porous nickel anode. This high temperature also allows for the usage of a wide range of fuels such as alcohols, hydrocarbons, and natural gas due to internal reforming of the fuel without the need for additional purifying or fuel processing. The last major benefit of operating at high temperatures is faster reactions as well as fast electrode kinetics, making SOFC highly efficient and providing high-quality waste heat for reuse. With high efficiency, this assists to lower operating costs and high energy production makes SOFCs ideal for industrial or commercial energy production. The current drawback of OXFCs are system controllability and thermal instability if any component have differing temperatures, but these can be mitigated through the implementation of a robust control system.

3.4.4 Phosphoric Acid

Phosphoric acid fuel cells (PAFCs) use a 100% pure concentration of phosphoric acid (H_3PO_4) as an electrolyte and operate at between 150 °C and 220 °C [13]. The matrix typically utilized to retain the acid is silicon carbide, as it is stable in the acid as well as a means for ion conduction. The highly concentrated phosphoric acid is highly stable and is a key contributor to the durability of the fuel cell and its operation at medium temperatures. As an elevated-temperature fuel cell, precious metals such as platinum must still be used as the electrocatalyst to assist the reactions at both electrodes. This high temperature allows for higher CO tolerance, about 1% CO, making it more robust and less sensitive to CO contamination compared to alkaline or PEM fuel cells.

In PAFCs hydrogen ions (H^+) are the medium of transfer between the anode and cathode from the electrochemical potential difference. With both the stability and robustness of PAFC, another benefit is high efficiency power generation of around 40–80% depending on if it is paired with a cogeneration system. Some of the drawbacks of PAFCs are relatively longer startup compared to other fuel cell, which may make it unideal for rapid deployment. PAFCs also suffer from volatilization of the phosphoric acid at higher operating temperatures, CO poisoning of the electrocatalyst, which may cause some instability and performance degradation. Still, PAFCs are still desirable in industrial applications due to their reliability, power efficiency, and relatively mature level of development compared to newer fuel cells.

3.4.5 Molten Carbonate

Molten carbonate fuel cells (MCFCs) use mixed alkali metal carbonates as an electrolyte in a porous ceramic matrix of $LiAlO_3$. To ensure the carbonates are molten, the typical operating temperature of an MCFC is between 600 °C and 700 °C allowing the use of nonprecious metals such as nickel (Ni) alloys, such as nickel aluminum as the anode and nickel oxide (NiO) as the cathode [14]. The high temperature allows for a wider range of fuels such as methane, natural gas, or even coal gas. The drawbacks of MCFCs are the high corrosiveness of the molten carbonate electrolyte, short life, and initial investment costs. Operationally, MCFCs require less continuous maintenance and have an efficiency around 60%.

3.5 Hybrid Systems

3.5.1 General Discussion

With assessment of both fuel cells and supercapacitors in the previous sections and batteries in other chapters makes it is evidently clear that each of them functions utilizing different principles to store electrical energy and as such has different benefits and drawbacks shown in Table 3.4. Ongoing research into these energy-storage technologies aims to close the gap between the systems though

Table 3.4 Battery vs supercapacitor vs fuel cell.

	Batteries	Supercapacitors	Fuel cells
Specific power (kW/kg)	1	10	0.9
Specific energy (Wh/kg)	250	10	1000
Efficiency	95%	98%	60%
Cyclic count	1000	200,000	4000
Usage	Transportation Backup power Portable electronics	Transportation	Industrial Grid
Benefits	Readily available energy and power Fast startup Mobile Low cost Mature technology Commercially available	Extremely long life High power delivery Mobile Cheap Mature technology Commercially available	Long life Negligible self-discharge Continuous energy
Drawbacks	Self-degradation Higher fire threat Finite capacity Commercially available	High self-discharge No long term storage large	Immobile Has startup time Fuel required Expensive Complex controls

advancements in materials and electrochemical research for energy and power densities. Fuel cells allow for the continuous production of clean power and can have high specific energy but require maintenance, costly, and relatively lower power density. Other benefits of fuel cells are their safety mechanisms, since they cannot produce energy if not turned on and their fuels are stored separately.

Supercapacitors have extremely high power density and quick response times but suffer from power energy density and inability to hold charge for extended periods of time. Supercapacitors are also more durable than either system allowing high currents without any damage to the system, some even up to 100 times greater than what would be safe for the other devices. Batteries are not the strongest in either specific power or specific energy, but their performance falls in the middle range for metrics. However, because their performance is nominal enough and most critically are the simplest to operate, batteries have become the most utilized technology of the three and are therefore most extensively researched.

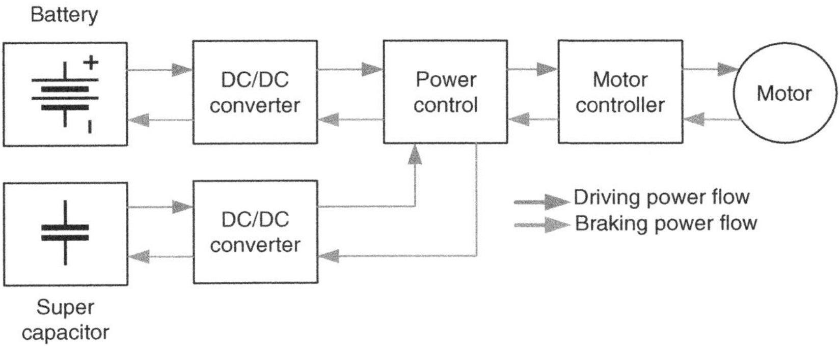

Figure 3.12 Vehicle hybrid battery fuel cell power flow.

3.5.2 Battery/Supercapacitor

Battery and supercapacitor hybrid systems greatly benefit from the addition of each part as the weaknesses of pure battery systems, such as low power density and slower reactivity, are well covered by the addition of supercapacitors. This combination is very well utilized in electrical systems requiring high power and constant power transfer, for example, electric vehicles (EVs) with regenerative braking. Since the high inrush current generated or supplied from the motor may exceed what the battery can do instantaneously, supercapacitors can act as the intermediary devices, managing the transient demand and supply from the motor while the battery system would cover the nominal steady-state operations. A basic power flow operation is shown below in Figure 3.12.

External to managing the main power flow, another example of a hybrid battery supercapacitor system can be shown with battery balancing. Battery balancing can be divided into two categories: active balancing and passive balancing. Active battery balancers aim to reduce state-of-charge variation or voltage differences between the cells. One very common approach is energy shuttling, where excess energy from the cell with the highest charge is transferred to an interim energy-storage device before shifting that energy to the cell with the lowest potential. This can be done on an individual level between adjacent cells in a stack, but in the case where the cells selected are not close to each other, the transferable energy would need to ripple through the cells not selected for transfer. Another method that requires less individual cell control is the use of a shared energy-storage tank, which could hold the transferred energy prior to shutting it to the lowest potential cell. This energy tank can suitably fit with a supercapacitor. Since the energy tank is a shared energy-transport device between many cells, the rate of power transfer and power delivered may be higher than what a typical capacitor could handle.

One such battery equalizer topology was proposed by Samanta et al. This battery balancing topology utilizes a supercapacitor alongside a bidirectional boost–buck

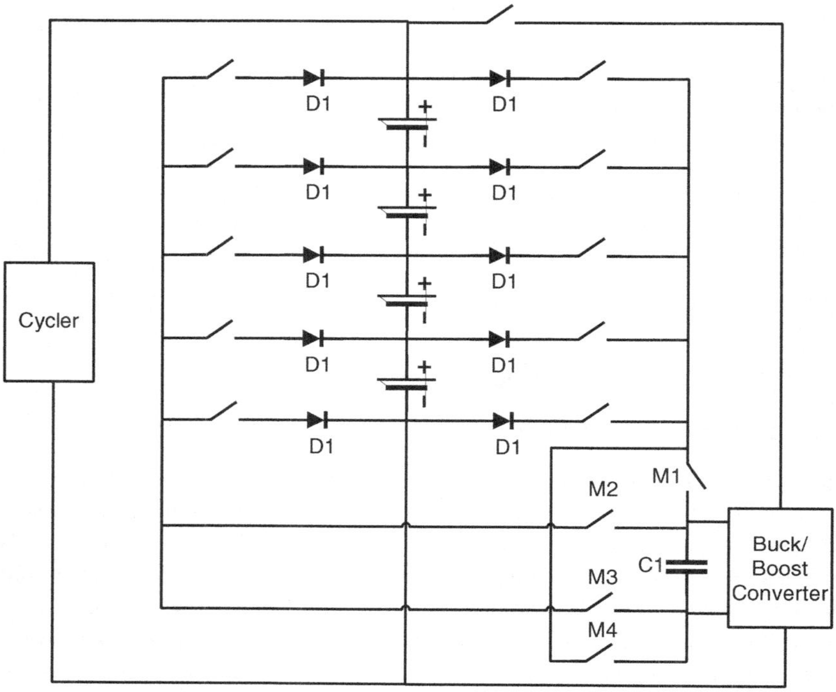

Figure 3.13 Battery balancer.

converter [16, 17]. It can work in three operating principles: cell-to-cell (C2C), cell-to-pack (C2P), and pack-to-cell (P2C) operations shown in Figure 3.13. For cell-to-cell operations, where only two cells are notably lower and higher than the pack average, the cell with the highest SOC could charge the supercapacitor and then redistribute the energy to the lower potential cell. Cell-to-pack operation is used for systems where a single cell is above average, taking energy to the supercapacitor and then boosting the energy to meet pack levels. Inversely, pack to cell would focus on the cell with the lowest potential, taking in overall pack energy and stepping down the voltage for the protected cell. Since all modes can be actively switched between utilizing the shared hardware system, the system undergoes high power transfer between the cells, this is typically small amounts of energy, but it could also store the energy for higher power delivery.

3.5.3 Battery/Fuel Cell

Fuel cell and battery hybrids are typically built to emulation of typical internal combustion generators or ICE/battery hybrid EVs. In older ICE/battery systems, there are many configurations for power delivery to the wheels, this could be

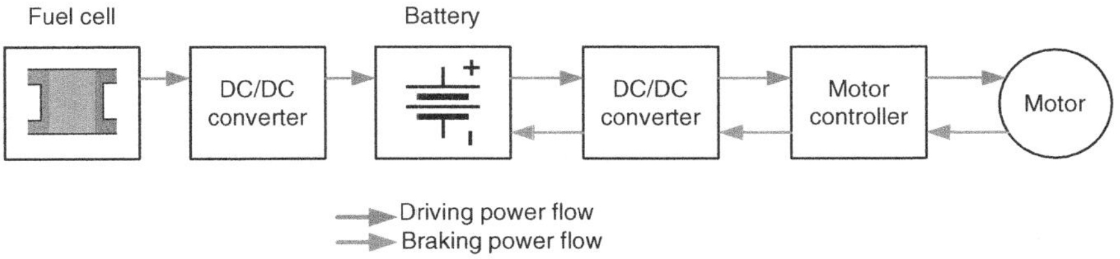

Figure 3.14 Fuel cell/battery hybrid charger.

achieved through parallel power usage of both the ICE and the battery through a planetary gearset. More modern EVs, with a greater reliance on the battery system, may use a smaller ICE to generate energy through a generator and recharge the batteries in a series configuration [18]. For fuel cells, the model can follow a series configuration since fuel cells lack the power required to directly drive the motor in the case of high transients shown in Figure 3.14. The power flow is primarily unidirectional, starting from the fuel cell, into the battery, then finally the load. Regenerative braking and energy recapture are possible, but this would be limited to the allowable operation and capacity of the battery, no energy should be sent backwards into the fuel cell as that would damage the fuel cell.

The battery act as a power and energy buffer, primarily dealing with the transient requirements of the vehicle. Secondary benefits of batteries for a fuel cell/battery vehicle include faster startup of the vehicle since fuel cells may need time to start up from dry or if the fuel cell needs time to manage the temperature or fuel. From a consumer perspective, the process and operation of an fuel cell/battery vehicle are already similar to those of an ICE vehicle, in which, if fuel is running low, the customer can refuel at a station or plug in the battery. Unfortunately, this is the main drawback of fuel cell vehicles as the infrastructure required for such convenience is not readily available, reducing confidence and increasing costs for early users of FCEVs.

3.6 Conclusion

Electrochemical energy is not limited to batteries alone, supercapacitors and fuel cells also have their applications and may often be paired with a battery to enhance a system performance. As such, some knowledge should be learned about complimentary energy-storage techniques and systems. Supercapacitors excel at power delivery and response but have poor long-term energy storage. Fuel cells have clean and continuous energy production, but are slower, low power, and cannot handle transient loads efficiently. When batteries, fuel cells, and supercapacitors are paired, the individual issues can be greatly mitigated but at a higher cost and complexity compared to a standalone system.

References

1 S. Ratha and A. K. Samantara, "*Springer Briefs in Materials Supercapacitor: Instrumentation, Measurement and Performance Evaluation Techniques*," (n.d.). http://www.springer.com/series/10111.

2 A. Kuperman and A. Lampasi, "*Advances in Supercapacitor Technology and Applications II*," (n.d.). www.mdpi.com/journal/energies.

3 K. K. Kar, "*Springer Series in Materials Science 313 Handbook of Nanocomposite Supercapacitor Materials III Selection*," (n.d.). http://www.springer.com/series/856.

4 K. K. Kar, (ed.), "*Handbook of Nanocomposite Supercapacitor Materials IV (Vol. 331)*," Springer International Publishing, 2023. https://doi.org/10.1007/978-3-031-23701-0.

5 A. Mendhe and H. S. Panda, "A review on electrolytes for supercapacitor device," Discov. mater., vol. 3, no. 1, 2023. https://doi.org/10.1007/s43939-023-00065-3.

6 P. Venet and E. Redondo-Iglesias, "*Batteries and Supercapacitors Aging*," (n.d.). www.mdpi.com/journal/batteries.

7 K. K. Kar, "*Springer Series in Materials Science 313 Handbook of Nanocomposite Supercapacitor Materials II Selection*," (n.d.). http://www.springer.com/series/856.

8 K. K. Kar, "*Springer Series in Materials Science 300 Handbook of Nanocomposite Supercapacitor Materials I Characteristics*," (n.d.). http://www.springer.com/series/856.

9 Y. Li, "*Fuel Cell Fundamentals and Applications*," 2025.

10 M. Butori, et al., "Ionic conductivity and hydrogen crossover for IT-PEMFCs: Influence of pressure, temperature, relative humidity and reinforcement," *Int. J. Hydrog. Energy*, 2024. https://doi.org/10.1016/j.ijhydene.2024.06.286., 2024

11 G. Huang, et al., "Composite Poly(norbornene) anion conducting membranes for achieving durability, water management and high power (3.4 W/cm2) in hydrogen/oxygen alkaline fuel cells," *J. Electrochem. Soc.*, vol. 166, no. 10, pp. F637–F644, 2019. https://doi.org/10.1149/2.1301910je.

12 J. Milewski, "*Advanced Methods of Solid Oxide Fuel Cell Modeling* (1st ed. 2011.)," Springer, 2011. https://doi.org/10.1007/978-0-85729-262-9.

13 P. S. Jiang and Q. Li, "*Introduction to Fuel Cells Electrochemistry and Materials Introduction to Fuel Cells Introduction to Fuel Cells*," 2022.

14 Z. Xu, "Porphyrin helical nanochannel-assembled polybenzimidazole membranes doped with phosphoric acid for fuel cells operating in a temperature range of 25–200 °C," *Adv. Funct. Mater.*, vol. 34 no. 12, 2024. https://doi.org/10.1002/adfm.202310762.

15 E. Açıkkalp, L. Chen, and M. H. Ahmadi, "Comparative performance analyses of molten carbonate fuel cell-alkali metal thermal to electric converter and molten carbonate fuel cell-thermo-electric generator hybrid systems," *Energy Rep.*, vol. 6, pp. 10–16, 2020. https://doi.org/10.1016/j.egyr.2019.11.108.

16 A. Samanta, A. Huynh, M. Sharma, V. Marcis, and S. Williamson, "Supercapacitor and bidirectional DC-DC converter-based active charge balancing scheme for lithium-ion batteries," *2022 IEEE Energy Conversion Congress and Exposition, ECCE 2022*, 2022. https://doi.org/10.1109/ECCE50734.2022.9947732

17 A. Samanta, M. Sharma, and S. Williamson, "A supercapacitor and Fuzzy-PID controller-based active charge balancing scheme for lithium-ion batteries," *IECON 2022 – 48th Annual Conference of the IEEE Industrial Electronics Society*, 2022. https://doi.org/10.1109/IECON49645.2022.9969110.

18 J. Bauman and M. Kazerani, "A comparative study of fuel-cell-battery, fuel-cell-ultracapacitor, and fuel-cell-battery-ultracapacitor vehicles," *IEEE Trans. Veh. Technol.*, vol. 57, no. 2, pp. 760–769, 2008. https://doi.org/10.1109/TVT.2007.906379.

4

Electrical Circuit Modeling, State Estimation, and Battery Pack Design

4.1 Introduction

Energy storage systems are fundamental to the operation of electric vehicles and modern electrified powertrains, where performance, safety, and reliability are governed by the dynamic behavior of lithium-ion batteries (LIBs) [1, 2]. Unlike conventional energy systems, LIBs exhibit highly nonlinear and time-varying characteristics influenced by operating conditions such as current load, temperature, and aging. As a result, accurate modeling, monitoring, and control of battery systems are essential for ensuring optimal performance and longevity [3]. From an operational perspective, a key challenge in battery systems is that critical internal states such as state of charge (SOC), state of health (SOH), and remaining useful life (RUL) cannot be measured directly. These states must be inferred from measurable parameters such as voltage, current, and temperature using mathematical models and algorithms [4]. Accurate state estimation is essential for reliable operation, optimal energy utilization, and safety assurance, particularly in electric vehicles where batteries operate under highly dynamic loading conditions [5]. For example, SOC estimation determines available driving range, while SOH and RUL estimation support maintenance planning and warranty decisions. Inaccurate estimation can lead to reduced usable capacity, performance degradation, and increased safety risks. Therefore, electrical modeling and state estimation techniques form the foundation of modern battery management systems (BMS), enabling real-time diagnostics, control, and prognostics.

In addition to the electrochemical behavior, thermal effects play a crucial role in battery performance and safety [6]. Heat is generated within the battery due to resistive losses and electrochemical reactions, leading to temperature rise and spatial gradients across cells and modules. Elevated ambient temperatures accelerate gradient mechanisms and increase the risk of thermal runaway, whereas low temperatures reduce power capability and efficiency. Consequently, battery thermal modeling and management are integral to battery systems to ensure battery operation within safe temperature limits and maintain uniform temperature distribution across the pack.

Furthermore, from a system-level perspective, a fundamental consideration in battery-powered systems is battery sizing, which directly influences vehicle range, performance, cost, weight, and safety. Battery sizing involves determining the required capacity, voltage, and current limits based on driving range, load profiles, and performance requirements, while ensuring operation within electrical, thermal, and safety constraints. These factors directly influence the selection of cell configuration, pack architecture, and cooling strategies. Unlike conventional power sources, battery systems must be sized not only for energy capacity but also for peak power demand, allowable depth of discharge, aging margins, and thermal limits. Improper sizing can result in reduced lifespan, insufficient power availability, or excessive system mass and cost. Therefore, battery sizing represents a critical early-stage design task that integrates system requirements, usage conditions, and battery characteristics into a coherent system-level design framework.

This chapter presents a comprehensive framework for battery system engineering, focusing on electrical circuit modeling, state estimation, thermal modeling, and battery pack design. It begins with battery sizing methodologies, followed by electrical equivalent circuit models used to represent battery dynamics. The chapter then discusses state estimation techniques for SOC, SOH, and RUL, along with their implementation in real-time systems. Thermal modeling and management strategies are subsequently introduced to address temperature regulation and safety. Finally, battery pack design considerations and system integration aspects are presented, highlighting their role in enabling efficient and reliable operation in practical applications.

4.2 Battery Sizing

Battery sizing is a foundational step in the design of energy storage systems for both transportation and grid-scale applications. Unlike conventional energy sources, batteries must simultaneously satisfy energy capacity, power capability, voltage constraints, life cycle requirements, and thermal limits. These requirements are governed by both cell-level characteristics and system-level operating conditions, which together determine achievable performance in practical applications. From a design perspective, battery sizing must ensure operation within a safe operating region defined by voltage, current, and temperature limits, while meeting performance and lifetime requirements. These constraints directly influence the selection of cell type, pack configuration, and thermal management strategy for a given application. Therefore, battery sizing establishes the critical link between cell-level characteristics and system-level design.

4.2.1 Key Parameters for Battery Sizing

The performance of LIBs is typically characterized by parameters such as specific energy, power density, capacity, cyclability, and safety limits. These parameters are influenced by operating conditions, including charge/discharge rate, temperature, voltage limits, SOC, and SOH [7]. Specific energy measured in Wh/kg specifies the energy a battery can store and release per unit mass. Specific capacity measures the amount of charge stored per unit mass and is expressed in ampere-hours per kilogram (Ah/kg). Specific energy is calculated as the product of the specific capacity and the operational battery voltage. Battery capacity is the amount of charge that can be drawn from a completely charged battery until it becomes completely discharged. Cyclability of LIBs is defined by the number of charge and discharge cycles until the battery reaches its end of life [8]. The charge/discharge rate is described by the C-rate, a magnitude measure of the current at which the batteries are charged/discharged. The safe operating region is characterized by voltage, current and temperature parameters. These parameters can be directly measured from the batteries. Considering both performance and safety parameters, the chemistry, size, and type of batteries are selected for a particular application [9]. This section begins with a detailed discussion of battery materials and components and their influence on energy and power density. These fundamentals form the basis for determining the number of cells, configuration, and overall size of battery systems used in electric vehicles as well as stationary grid applications [10].

4.2.1.1 Electrode Materials, Cell Voltage, and Cell sizing

The voltage of a LIB cell is determined by the electrochemical potentials of the anode and cathode materials. In LIBs, energy storage is based on the reversible intercalation and deintercalation of lithium ions between the two electrodes [11]. The potential of each electrode is defined relative to a reference, typically lithium metal, which is assigned to a potential of 0 V. When an electrode material is paired with lithium metal in a half-cell configuration, the measured voltage represents its half-cell potential, which varies with lithium concentration in the electrode [12]. For example, graphite anodes exhibit a low and relatively flat potential versus lithium during lithium ion intercalation into the graphite structure, while common cathode materials operate at significantly higher potentials. These half-cell potentials are strongly dependent on the electrode material and SOC. In a full lithium-ion cell, the cell voltage is given by the difference between the cathode and anode potential:

$$V_{\text{cell}} = U_{\text{cathode}} - U_{\text{anode}}$$

As lithium ions move between the electrodes during charge and discharge, the individual electrode potentials change, leading to a corresponding variation in cell voltage. The nominal cell voltage (typically 3.2–3.7 V) represents an average operating value over the usable SOC range rather than a fixed quantity. From a battery sizing perspective, the nominal voltage determines the number of cells required in series to achieve a desired system voltage. The selection of electrode materials, therefore, directly impacts voltage level, energy density, and allowable operating limits, all of which are critical considerations in the design of battery systems for electric vehicles and grid energy storage applications.

Example: A Li-ion cell uses:

- Graphite anode with an average half-cell potential $\bar{U}_- = 0.12$ V vs. Li/Li$^+$
- NMC cathode with an average half-cell potential $\bar{U}_+ = 3.85$ V vs. Li/Li$^+$

Part A (nominal voltage method): Determine the nominal full-cell voltage using the average half-cell potential.

Part B: A cell has:

- Cathode capacity $C_+ = 2.0$ Ah
- Anode capacity $C_- = 2.6$ Ah
- Total cyclable Li $Q_{\text{Li}} = 1.8$ Ah

Compute the open-circuit voltage when 45% of Li is in the anode using half-cell OCV curves.

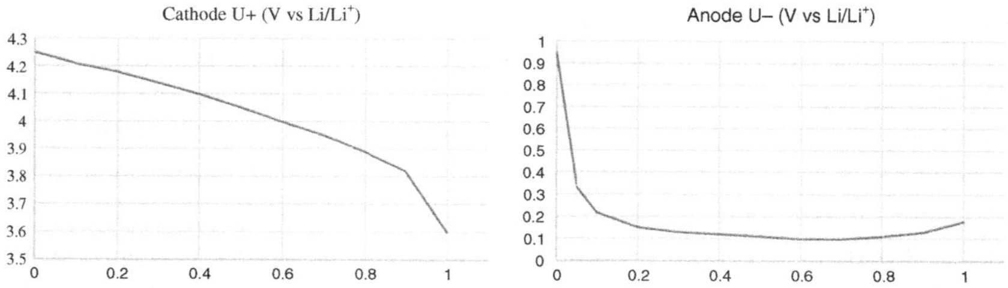

Solution:

Part A: Nominal voltage from average half-cell potential

Nominal cell voltage: $V_{\text{nom}} = \bar{U}_+ - \bar{U}_- = 3.85 - 0.12 = 3.73$ V

$$V_{\text{nom}} \approx 3.73 \text{ V}$$

For preliminary design and sizing purposes, the cell voltage is often approximated using the average half-cell potential over the intended operating SOC window. This approach neglects the variation of electrode potentials with Li stoichiometry and therefore does not capture the SOC-dependent nature of the open-circuit voltage. To account for the dependence of cell voltage on Li distribution between electrodes, a curve-based approach using half-cell open-circuit potential functions is employed.

Part B: OCV based

The Li distribution is converted into electrode stoichiometries by normalizing the Li content with respect to the respective electrode capacities.

Use:

$$U_{\text{cell}} = U_{+}(y) - U_{-}(x)$$

Step 1: Li split (45% in anode)

$$Q_{\text{Li,anode}} = 0.45 \times 1.8 = 0.81 \text{ Ah}$$

$$Q_{\text{Li,cathode}} = 1.8 - 0.81 = 0.99 \text{ Ah}$$

Step 2: Convert to stoichiometries x and y

$$x = \frac{Q_{\text{Li,anode}}}{C_{-}} = \frac{0.81}{2.6} = 0.312$$

$$y = \frac{Q_{\text{Li,cathode}}}{C_{+}} = \frac{0.99}{2.0} = 0.495$$

Step 3: Read potentials from the half-cell curves (or interpolate)
From the provided dataset below (linear interpolation is fine):

$$U_{+}(y \approx 0.495) \approx 4.05 \text{ V}$$

$$U_{-}(x \approx 0.312) \approx 0.12 \text{ V}$$

So: $U_{\text{cell}} \approx 4.05 - 0.12 = 3.93$ V

$$U_{\text{cell}} \approx 3.93 \text{ V (OCV at this Li distribution)}$$

Together, Parts A and B illustrate the difference between nominal voltage estimation for system-level design and detailed voltage evaluation based on electrode thermodynamics.

- Comment: How electrode selection influences system voltage design.
- Higher cathode potential (e.g., high-Ni NMC) or lower anode potential increases cell voltage, which reduces the required series cell count for a given pack voltage.
- However, electrode choices also affect safety margin, degradation, and allowable voltage windows, so system voltage design must consider not only nominal voltage but also min/max voltage, thermal behavior, and aging.

4.2.1.2 Specific Capacity, Ah vs. Wh, and Energy Density

While electrode materials determine the cell voltage, the amount of energy that can be stored in a battery cell is governed by its capacity. Battery capacity is commonly expressed in ampere-hours (Ah), whereas the usable energy content is expressed in watt-hours (Wh). Understanding the distinction between these quantities, and their relationship to material properties and cell voltage, is essential for accurate battery sizing and comparison across technologies.

Specific Capacity (Ah): The specific capacity of a battery electrode refers to the amount of electric charge that can be stored per unit mass of active material and is typically expressed in mAh/g. At the cell level, the rated capacity C (in Ah) represents the total charge that can be delivered when the battery is discharged from its upper to lower voltage limits under specified conditions. For a battery cell operating at a current I for a duration t, the capacity is defined as:

$$C = \int_0^t I(\tau)\,d\tau$$

In practice, manufacturers specify the nominal capacity based on a standardized discharge rate (e.g., C/10 or C/5), as the available capacity decreases at higher current rates due to polarization and transport limitations. Importantly, the usable cell capacity is limited by electrode with the lower reversible Li storage capability, making electrode balancing a critical design consideration.

Ampere-hours (Ah) versus Watt-hours (Wh): Although ampere hours quantify the amount of charge stored, they do not directly represent the energy content of a battery. The actual energy delivered depends on the operating voltage of the cell. The energy E stored in a battery is given by:

$$E = \int V(t)\,I(t)\,dt$$

For practical calculations, this is commonly approximated using the nominal voltage V_{nom}:

$$E \approx V_{\text{nom}} \times C$$

Thus, two battery cells with the same Ah capacity but different nominal voltages will store different amounts of energy. This distinction is critical when comparing batteries based solely on capacity ratings, as Ah values alone do not capture differences arising from electrode chemistry or voltage characteristics.

Specific Energy and Energy Density: The specific energy of a battery cell is defined as the energy stored per unit mass and is expressed in Wh/kg

$$E_{\text{specific}} = \frac{E}{m}$$

where m is the total mass of the cell includes both active and inactive components such as current collectors, separators, electrolyte, and packaging. Similarly, volumetric energy density (Wh/L) represents the energy stored per unit volume and is particularly important in applications where space is constrained. While electrode materials with high theoretical specific capacity and high operating voltage contribute to increased specific energy, the practical values achieved at the cell level are significantly lower due to inactive materials and safety margins. Consequently, system-level battery sizing must always rely on cell-level specific energy values rather than theoretical material limits.

Example: Two LIB cells have the same rated capacity of 4 Ah, but different nominal voltages:

- Cell A: $V_{\text{nom}} = 3.2$ V
- Cell B: $V_{\text{nom}} = 3.8$ V

Both cells have a mass of 0.48 kg.

1. Determine the energy stored in each cell in Wh.
2. Calculate the specific energy (Wh/kg) of each cell.
3. Comment on why Ah alone is insufficient to compare battery energy content.

Solution:

Step 1: Energy content

$$E = V_{\text{nom}} \times C$$

Cell A: $E_{\text{A}} = 3.2 \times 4 = 12.8$ Wh
Cell B: $E_{\text{B}} = 3.8 \times 4 = 15.2$ Wh

Step 2: Specific energy

$$E_{\text{specific}} = \frac{E}{m}$$

Cell A: $E_{\text{specific,A}} = \frac{12.8}{0.48} = 26.7$ Wh/kg

Cell B: $E_{\text{specific,B}} = \frac{15.2}{0.48} = 31.7$ Wh/kg

Although both cells have the same ampere-hour capacity, Cell B stores more energy due to its higher nominal voltage. This example demonstrates that ampere-hour ratings alone do not reflect the true energy content of a battery.

4.2.1.3 C-rate, Power Capability, and Rate-dependent Capacity

While specific capacity and energy density define how much charge and energy a battery can store, the rate at which this energy can be delivered or absorbed is equally important. This rate capability is commonly expressed using the C-rate, which links battery capacity to operating current and directly influences power capability, efficiency, and usable capacity.

Definition of C-Rate: The C-rate is a normalized measure of charge or discharge current relative to the rated capacity of a battery. For a battery with nominal capacity $C(\text{Ah})$, the current corresponding to a C-rate of nC is given by:

$$I = n \times C$$

For example, a 5 Ah cell discharged at:

- 1C delivers 5 A and would ideally discharge in 1 hour,
- 0.5C delivers 2.5 A and would discharge in 2 hours,
- 2C delivers 10 A and would be discharged in 0.5 hours.

C-rate provides a convenient, chemistry-independent way to compare operating conditions across batteries of different sizes.

Power Capability and Internal Resistance: The instantaneous electrical power delivered by a battery is given by:

$$P = V \times I$$

At higher C-rates, the operating current increases, enabling higher power output. However, increased current also leads to larger voltage drops due to the internal resistance of the cell, reducing the effective terminal voltage. As a result, the maximum usable power of a battery is constrained not only by capacity but also by internal resistance and allowable voltage limits. Cells designed for high-power operation typically feature thinner electrodes, higher porosity, and enhanced electronic and ionic conductivity. These design choices improve rate capability but reduce the amount of active material per unit volume, leading to a trade-off between specific energy and specific power.

Rate-Dependent Capacity: The usable capacity of a battery decreases as the discharge rate increases, a phenomenon known as rate-dependent capacity loss. At high C-rates, limitations in Li-ion transport, electrode kinetics, and electrolyte conductivity prevent full utilization of the active material within the allowed voltage window. Consequently, the measured capacity at high discharge rates is lower than the nominal capacity specified at low rates. This effect is particularly important when interpreting ampere-hour ratings, as capacity values provided by manufacturers are typically measured under standardized, low-rate conditions. In practical operation, especially under dynamic loading, the available capacity may be significantly lower.

Implications for Battery Sizing: The concepts of C-rate and rate-dependent capacity establish that battery sizing cannot rely solely on nominal capacity and energy values. Instead, the intended operating current range and power demand must be considered to ensure that sufficient capacity remains available under real operating conditions. These considerations become increasingly important in applications requiring frequent high-power transients or rapid charge-discharge cycles.

4.2.2 Battery Capacity and Power Sizing Methodology

Battery sizing is the process of determining the required energy capacity, power capability, and electrical configuration of a battery system such that it can reliably meet application demands while operating within safe voltage, current, and temperature limits. Building on the concepts of cell voltage, capacity, energy, and C-rate, this section presents a systematic methodology for sizing battery systems at the cell and pack levels.

4.2.2.1 Energy-based Sizing

The first step in battery sizing is determining the energy requirement of the application. The required usable energy E_{req} is typically specified based on operational needs over a given duration. At the battery level, the usable energy is related to the nominal battery energy by accounting for allowable DOD:

$$E_{\mathrm{usable}} = \mathrm{DOD} \times E_{\mathrm{nominal}}$$

Rearranging, the nominal energy requirement becomes:

$$E_{\mathrm{nominal}} = \frac{E_{\mathrm{req}}}{\mathrm{DOD}}$$

Since battery cells are rated in ampere-hours, the required pack capacity C_{pack} (Ah) can be expressed as:

$$C_{\mathrm{pack}} = \frac{E_{\mathrm{nominal}}}{V_{\mathrm{pack,nom}}}$$

where $V_{\mathrm{pack,nom}}$ is the nominal pack voltage determined from the series connection of cells.

4.2.2.2 Voltage Determination and Series Cell Configuration

The nominal pack voltage is obtained by connecting cells in series:

$$V_{\mathrm{pack,nom}} = N_s \times V_{\mathrm{cell,nom}}$$

where N_s is the number of cells in series, $V_{\mathrm{cell,nom}}$ is the nominal voltage of a single cell.

The value of N_s is chosen to satisfy system voltage requirements while ensuring that:

- maximum cell voltage does not exceed upper safety limits during charging,
- minimum cell voltage remains above cutoff limits during discharge.

Voltage selection therefore directly links electrode material choice (Section 4.2.1.1) with system-level electrical architecture.

4.2.2.3 Capacity Scaling and Parallel Cell Configuration

Once the series cell count is established, the required capacity is achieved by connecting multiple cells in parallel. The total pack capacity is given by:

$$C_{\text{pack}} = N_{\text{p}} \times C_{\text{cell}}$$

where N_{p} is the number of parallel cells, C_{cell} is the nominal capacity of a single cell.

The number of parallel cells is therefore:

$$N_{\text{p}} = \frac{C_{\text{pack}}}{C_{\text{cell}}}$$

Parallelization increases total capacity and reduces current per cell, improving thermal performance and cycle life.

Example: A stationary energy storage system is required to deliver $E_{\text{req}} = 2.5$ MWh of usable energy per event. To improve cycle life, the system is operated at a maximum DOD of 80%. The inverter DC bus is designed for a nominal battery voltage of $V_{\text{pack,nom}} = 720$ V.

Battery cells available:

- Nominal cell voltage: $V_{\text{cell,nom}} = 3.6$ V
- Nominal cell capacity: $C_{\text{cell}} = 280$ Ah

Determine:

1. The required nominal battery energy (MWh)
2. The required pack capacity (Ah)
3. The required number of cells in series N_s
4. The required number of parallel strings N_p
5. The final pack configuration $N_s \times N_p$

Solution: Usable energy is related to nominal energy by: $E_{\text{usable}} = \text{DOD} \times E_{\text{nominal}}$

$$E_{\text{nominal}} = \frac{E_{\text{req}}}{\text{DOD}} = \frac{2.5}{0.8} = 3.125 \text{ MWh}$$

Step 1: Compute required pack capacity (Ah)

Using the approximation: $E \approx V_{\text{pack,nom}} \times C_{\text{pack}}$

$$C_{\text{pack}} = \frac{E_{\text{nominal}}}{V_{\text{pack,nom}}} = \frac{3.125 \times 10^6}{720} = 4340.28 \text{ Ah}$$

Step 2: Determine the number of series cells

$$N_s = \frac{V_{\text{pack,nom}}}{V_{\text{cell,nom}}} = \frac{720}{3.6} = 200$$

Step 3: Determine the number of parallel strings

$$C_{\text{pack}} = N_p \times C_{\text{cell}} \Rightarrow N_p = \frac{C_{\text{pack}}}{C_{\text{cell}}} = \frac{4340}{280} = 15.5 \approx 16$$

Resulting pack capacity: $C_{\text{pack,actual}} = 16 \times 280 = 4480$ Ah

Step 4: Check energy delivered with this configuration

Nominal pack energy:

$$E_{\text{nom,actual}} = V_{\text{pack,nom}} \times C_{\text{pack,actual}} = 720 \times 4480 = 3{,}225{,}600 \text{ Wh} = 3.22 \text{ MWh}$$

Usable energy at 80% DOD:

$$E_{\text{usable,actual}} = 0.8 \times 3.226 = 2.581 \text{ MWh}$$

Hence,

- The total number of cells $= N_s \times N_p = 200 \times 16$
- Nominal pack voltage: $200 \times 3.6 = 720$ V
- Pack capacity: $16 \times 280 = 4480$ Ah
- Usable energy (80% DOD): ≈ 2.58 MWh

4.2.2.4 Power-based Sizing and C-rate Constraints

Energy-based sizing alone is insufficient if the battery must deliver or absorb high power. The maximum allowable current per cell is constrained by the permissible C-rate:

$$I_{\text{cell,max}} = C_{\text{cell}} \times C_{\text{rate,max}}$$

The total pack current capability is then:

$$I_{\text{pack,max}} = N_p \times I_{\text{cell,max}}$$

Accordingly, the maximum pack power is:

$$P_{\text{pack,max}} = V_{\text{pack}} \times I_{\text{pack,max}}$$

If the required power exceeds this limit, the number of parallel cells must be increased, even if the energy requirement is already satisfied. This highlights the need to consider both energy and power constraints simultaneously during sizing.

In practice, battery sizing must satisfy both of the following conditions:

$$E_{\text{pack}} \geq E_{\text{required}}$$
$$P_{\text{pack,max}} \geq P_{\text{required}}$$

The final configuration (N_s, N_p) is selected as the minimum combination that satisfies both criteria while remaining within safe operating limits for voltage, current, and temperature. Furthermore, design margins and practical considerations: To ensure reliable operation over the battery lifetime, sizing calculations typically include additional design margins to account for capacity fading due to aging, temperature-dependent performance variations, manufacturing tolerances and cell imbalance, future operational uncertainty. As a result, the final battery size is often larger than the theoretical minimum required to meet initial energy and power demands.

4.3 LIB States

The parameters that are essential for monitoring and managing the performance and lifespan of LIBs in various applications are shown in Figure 4.1, including SOC, SOH, state of power (SOP), and state of temperature (SOT). In addition, the derived estimators include state of energy (SOE), depth of discharge (DOD), and remaining useful life (RUL). However, these states need to be estimated and cannot be measured directly from the battery using the operating parameters of the battery such as voltage, temperature, and current.

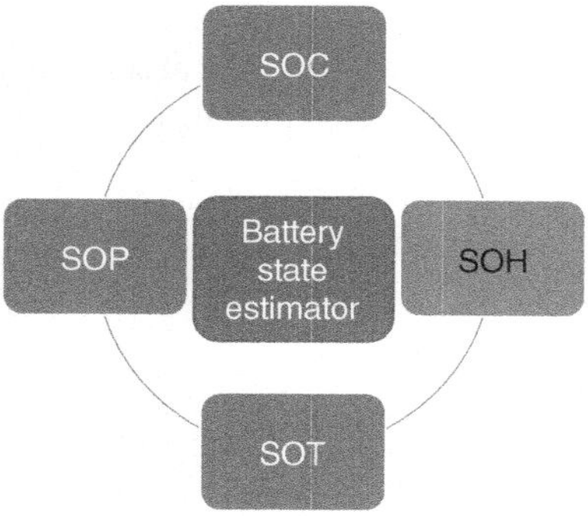

Figure 4.1 States of LIBs.

4.3.1 State of charge

Among various battery states, SOC estimation is a crucial requirement for a BMS. It ensures the operation of the battery pack within the safe operating area (SOA) as defined by the LIB pack manufacturer. Additionally, SOC estimation provides information on the remaining useful energy and vehicle driving range, similar to the fuel gauge of an internal combustion engine [7, 13–15]. It ensures efficient use of capacity to maximize the

EV range and achieve better cycle life. Accurate SOC estimation also helps prevent overcharging and over-discharging of batteries, which are potential safety hazards associated with LIBs. The fundamental definition of SOC is the ratio of the available battery capacity ($Q_{\text{available}}$) to its maximum rated capacity (Q_{Rated}) [16]. The mathematical representation of the definition is shown in (4.1).

$$\text{SOC} = \frac{Q_{\text{available}}}{Q_{\text{rated}}} \tag{4.1}$$

Accurate SOC estimation of LIBs is a complex and crucial aspect of BMS design, despite the seemingly simple equation used to calculate SOC. Practically, SOC cannot be accurately estimated solely based on battery capacity as per the theoretical definition described in (4.1). SOC estimation enables enforcement of voltage and current limits, prevention of overcharge and over-discharge, optimal energy utilization, and reliable range or energy availability prediction. However, SOC cannot be directly measured and must be inferred from measurable signals such as current, voltage, and temperature using estimation algorithms. Over the past decades, numerous SOC estimation methods have been developed. These methods can be broadly classified into direct measurement-based methods, model-based methods, and data-driven methods as shown in Figure 4.2, each offering distinct advantages and limitations in terms of accuracy, robustness, and computational complexity.

4.3.1.1 Direct Methods

a. Coulomb Counting Method
The SOC of a LIB using capacity is computed through the Coulomb counting method shown in (4.2), by counting ampere hour and integrating the battery current over a period.

$$\text{SOC}(t) = \text{SOC}(t_0) + \frac{\eta}{Q_{\text{Rated}}} \int_{t_0}^{t} I_{\text{batt}} dt \tag{4.2}$$

where $\text{SOC}(t_0)$ is the initial SOC value, η is the Coulombic efficiency, Q_{Rated} is the maximum rated battery capacity, I_{batt} is the battery current, and $\text{SOC}(t)$ is the present battery SOC. This method is widely used to estimate SOC in EVs. However, the accuracy is dependent on initial SOC value, $\text{SOC}(t_0)$, open circuit voltage (OCV), temperature, battery internal resistance, and battery age, SOH, and chemistry.

b. OCV-based Estimation
OCV-based SOC estimation exploits the monotonic relationship between the battery open-circuit voltage and SOC under equilibrium conditions. By measuring the battery voltage after sufficient rest and referencing an experimentally obtained OCV-SOC curve, SOC can be inferred.

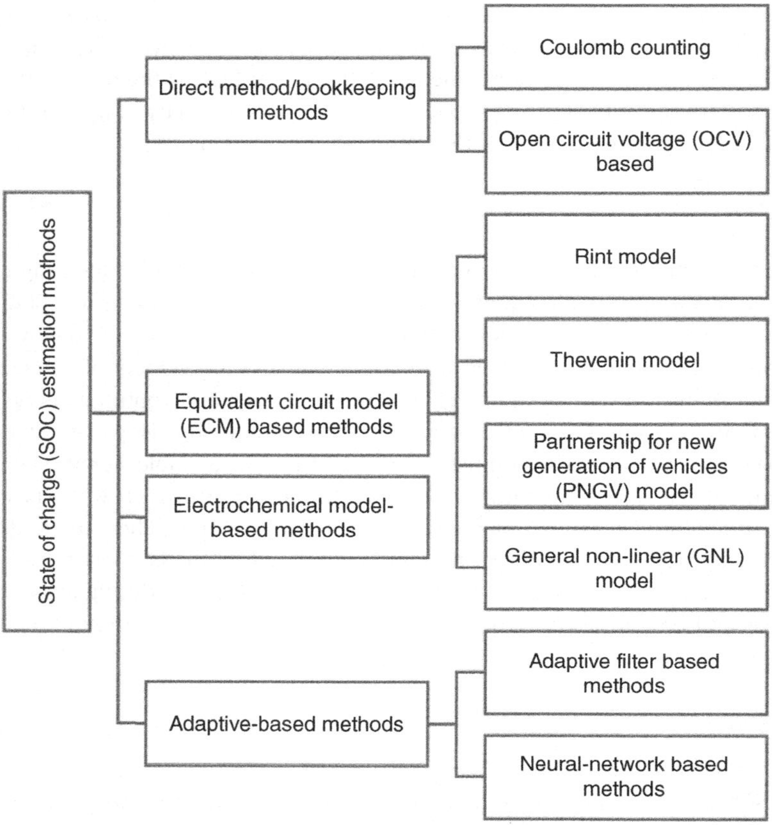

Figure 4.2 SOC estimation methods.

$$\text{SOC} = f^{-1}(V_{\text{OCV}})$$

Advantages:

- High accuracy under equilibrium conditions
- No cumulative error

Limitations:

- Requires long rest periods
- OCV-SOC relationship is chemistry- and temperature-dependent
- Not suitable for dynamic operating conditions

OCV-based estimation is commonly used to initialize SOC or to periodically correct Coulomb counting drift.

Examples:

1. SOC from an OCV: SOC Lookup Table below, a Li-ion cell has the following 25 °C OCV-SOC table, after resting, the measured open-circuit voltage is 3.68 V. Estimate SOC.

SOC (%)	OCV (V)
0	3.00
10	3.30
20	3.42
30	3.52
40	3.60
50	3.68
60	3.75
70	3.83
80	3.92
90	4.05
100	4.20

Solution: From the table, 3.68 V corresponds exactly to 50% SOC.

2. Using the same 25 °C table above, a rested cell has OCV = 3.78 V. Estimate SOC using linear interpolation.

Solution: Locate the interval containing 3.78 V:

- At 60% SOC, OCV = 3.75 V
- At 70% SOC, OCV = 3.83 V

So, SOC lies between 60% and 70%.
Linear interpolation:

$$\text{SOC} = 60 + \frac{(3.78 - 3.75)}{(3.83 - 3.75)} \times (70 - 60)$$

$$\text{SOC} = 60 + 0.375 \times 10 = 63.75\%$$

3. A cell has the following OCV-SOC data at 0 °C and 25 °C:

0 °C		25 °C	
SOC (%)	OCV (V)	SOC (%)	OCV (V)
40	3.58	40	3.60
50	3.65	50	3.68

(Continued)

0 °C		25 °C	
SOC (%)	OCV (V)	SOC (%)	OCV (V)
60	3.72	60	3.75
70	3.80	70	3.83

A rested cell at 10 °C has OCV = 3.70 V. Estimate SOC at different temperatures using two-table interpolation:

1. SOC from the 0 °C table
2. SOC from the 25 °C table
3. Interpolate SOC to 10 °C

Solution:

Step 1: SOC estimate using 0 °C table

At 0 °C:

- 50% SOC corresponds to 3.65 V
- 60% SOC corresponds to 3.72 V

Interpolate:

$$\begin{aligned}\mathrm{SOC}_0 &= 50 + \frac{(3.70 - 3.65)}{(3.72 - 3.65)} \times 10 \\ &= 50 + \frac{0.05}{0.07} \times 10 = 50 + 7.1429 = 57.14\%\end{aligned}$$

Step 2: SOC estimate using 25 °C table

At 25 °C:

- 50% SOC corresponds to 3.68 V
- 60% SOC corresponds to 3.75 V

$$\begin{aligned}\mathrm{SOC}_{25} &= 50 + \frac{(3.70 - 3.68)}{(3.75 - 3.68)} \times 10 \\ &= 50 + \frac{0.02}{0.07} \times 10 = 50 + 2.857 = 52.86\%\end{aligned}$$

Step 3: Interpolate SOC to 10 °C

Temperature fraction between 0 °C and 25 °C:

$$\alpha = \frac{10 - 0}{25 - 0} = 0.4$$

$$\begin{aligned}\mathrm{SOC}_{10} &= \mathrm{SOC}_0 + \alpha(\mathrm{SOC}_{25} - \mathrm{SOC}_0) \\ &= 57.14 + 0.4(52.86 - 57.14) \\ &= 57.14 + 0.4(-4.28) = 57.14 - 1.71 \\ &= 55.43\%\end{aligned}$$

4.3.1.2 Model-based Estimation Methods

Model-based SOC estimation techniques use mathematical battery models to relate measurable inputs (current, voltage, temperature) to internal states. These methods are particularly effective under dynamic operating conditions.

a. Equivalent Circuit Model (ECM)-based Methods

ECMs are descriptive models that replicate the battery charge–discharge behavior. The ECM of a LIB is built using electronic circuits like resistors, parallel pairs of resistors, and capacitors. This is because the behavior and performance of the batteries are analogous to the properties and behavior of electronic circuit components. The ECM models can be categorized as the integral-order (IOM) models and fractional-order models (FOM) [17]. IOMs are time domain analysis models and use few parameters to represent the relationship between voltage and current. The most common time domain analysis-based models are, as shown in Figure 4.3, the Rint model, Thevenin model, partnership for a new generation of vehicle (PNGV) model, and general non-linear model (GNL) [18]. An ECM model characterizes the behavior of the batteries incorporating voltage sources and electric circuit components. The model parameters are obtained by parameter identification, which involves experimental data such as voltage, current and temperatures during the battery charge and discharge process.

ECM-based SOC estimation methods involve the usage of the model to simulate battery behavior and evaluate the predicted output from simulation to the actual measurement results from the real battery. However, the model prediction deviates from the true system response due to noises in the sensor readings and erroneous model parameters.

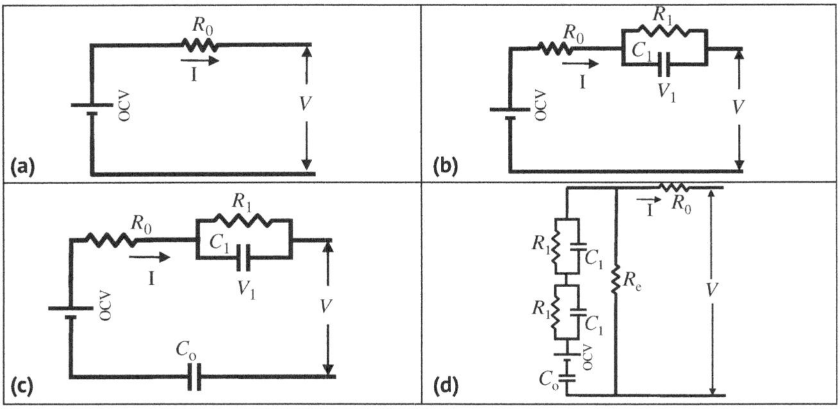

Figure 4.3 Equivalent circuit models: (a) Rint, (b) Thevenin, (c) PNGV, and (d) GNL.

b. Electrochemical Model-based Methods

Electrochemical models represent battery behavior by capturing the intrinsic physical states of the cell. These physics-based models are typically formulated using partial differential equations (PDEs), which describe the transport phenomena and electrochemical reactions within the battery. However, the use of PDEs leads to high computational complexity, making them challenging for real-time applications. To address this, PDEs are often reduced to ordinary differential equations (ODEs) through discretization techniques. Common numerical methods employed for this purpose include integral approximation, Padé approximation, the Ritz method, as well as finite element and finite difference methods [18].

Among the widely studied physics-based models, the single particle model (SPM) and the pseudo-two-dimensional (P2D) model are the most prominent [19]. The SPM is a simplified electrochemical model that represents Li-ion concentration and diffusion within electrode particles. While it offers reduced computational complexity and is suitable for practical implementation, its accuracy is limited due to the simplification of electrochemical processes, particularly under high-rate operating conditions. To improve its performance, extended versions of the SPM incorporate effects such as non-uniform reaction distribution and electrolyte dynamics [20].

In contrast, the P2D model provides a more comprehensive representation by incorporating porous electrode theory, electrolyte concentration dynamics, and reaction kinetics to describe battery behavior. Although the P2D model achieves higher accuracy, it is computationally intensive, which limits its direct application in real-time systems. Consequently, various reduced-order and simplified P2D models have been developed to balance accuracy and computational efficiency [17].

Advantages:

- Strong physical interpretability
- High accuracy

Limitations:

- High computational cost
- Limited suitability for embedded BMS applications

4.3.1.3 Adaptive Methods

a. Adaptive Filter-based Methods

Adaptive filter-based SOC estimation methods are widely used in battery management systems due to their ability to handle model uncertainties and measurement noise in real time. The prominent filtering approaches utilized in LIB applications are the Kalman filter (KF) and particle filter (PF), and their variants, including extended Kalman filter (EKF) unscented Kalman filter (UKF), and unscented particle filter (UPF). Nevertheless, the battery model needs to be precise for the filtering method to be effective. To maintain the SOC estimation accuracy throughout the cycle life of the battery, the model needs to be highly adaptive to the dynamic behavior of a battery under different operating conditions such as temperature, current loading, and capacity degradation due to aging. KF is a recursive algorithm best suited for linear systems in solving state estimation issues. The KF works with the assumption that all noises are white and Gaussian. The recursiveness is enabled by predicting the states and their uncertainties and then correcting these predictions using the measured values. A KF can be visualized as shown in Figure 4.4.

The discrete-time model for SOC is given by (4.1).

$$z[k+1] = z[k] - \frac{\Delta t}{Q}\eta[k]i[k] \tag{4.1}$$

Here, SOC is represented as $z[k]$, Δt is sampling time, Q is the total capacity, η is the columbic efficiency and i is the current through the battery. The terminal voltage is modeled by the equation (4.2).

$$v[k] = \mathrm{OCV}(z[k]) - i[k]R_s \tag{4.2}$$

Considering the RC parallel circuit, the expression for terminal voltage is modeled as (4.3).

$$Z[k] = \mathrm{OCV}(z[k]) - i_{R_{ct}}[k]R_{ct} - i[k]R_s \tag{4.3}$$

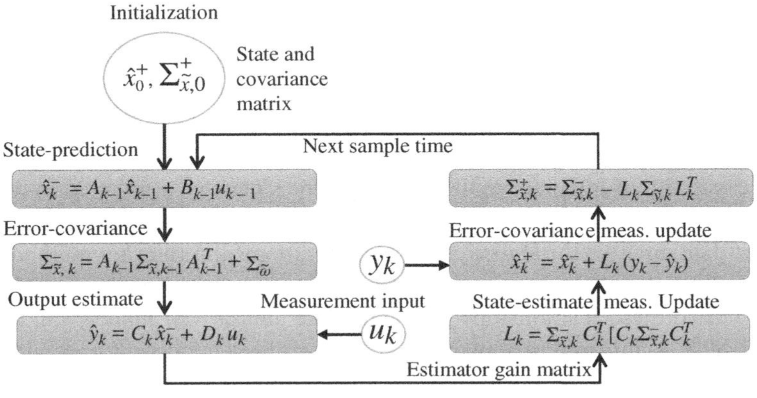

Figure 4.4 Kalman filter realization.

The current flowing through the parallel RC branch is modeled in the discrete time as (4.4).

$$i_{R_{ct}}[k+1] = \exp\left(-\frac{\Delta t}{R_{ct}C_{dl}}\right) \times i_{R_{ct}}[k] + \left(1 - \exp\left(-\frac{\Delta t}{R_{ct}C_{dl}}\right)\right) \times i[k] \tag{4.4}$$

If multiple RC branches are used (tau = exp $(-\Delta t/Rj\,Cj)$, where $j = 1, 2, 3 \ldots .$) to model the ECM, then the state equation for resistor currents can be written as (4.5).

$$i_R[k+1] = \begin{bmatrix} \tau_1 & 0 & \ldots \\ 0 & \tau_2 & \cdots \\ \vdots & \vdots & \ddots \end{bmatrix} i_{R,k} + \left(i_{R,k} + \omega_k\right) \tag{4.5}$$

The model output equation is given by (4.6).

$$y_k = \mathrm{OCV}(z_k) + Mh_k + M_0 s_k - \sum R_j i_{j,k} - R_s i_k + v_k \tag{4.6}$$

Here, M and M0 in the above equation correspond to hysteresis phenomena that appear with the charge-discharge process.

Advantages:

- Good accuracy under dynamic conditions
- Real-time implementation feasible
- Can incorporate temperature and aging effects

Limitations:

- Requires accurate model parameterization
- Performance degrades with model mismatch

b. Data-Driven and Machine Learning Methods

Data-driven models have been increasingly employed to accurately capture the nonlinear behavior of LIBs. These approaches utilize machine learning techniques to establish relationships between measurable signals, such as voltage, current, and temperature, and internal battery states using large datasets. A key advantage of data-driven methods is their ability to learn and adapt to changes in battery behavior over time, including aging effects. Their capability to integrate multiple data sources enables a more comprehensive representation of battery dynamics. Model-based techniques rely on indirect data analysis through complex computations and filtering methods to estimate battery parameters [21]. In contrast, data-driven approaches directly learn these relationships from data. The data-driven models can be categorized [22] into neural networks, Gaussian process regression, ensemble learning, transfer learning, and conventional methods. In addition, techniques such as fuzzy logic, support vector machines, and decision trees have also been widely applied for LIB modeling.

Advantages:

- Capable of modeling nonlinear behavior
- Reduced dependence on explicit battery models

Limitations:

- Require large, representative training datasets
- Limited extrapolation capability
- Reduced physical interpretability

4.3.1.4 Hybrid SOC Estimation Approaches

Hybrid SOC estimation methods combine the strengths of model-based and data-driven approaches to improve accuracy and robustness under varying operating conditions [23]. Typically, these methods integrate equivalent circuit or electrochemical models with machine learning algorithms or adaptive filters to leverage both physical interpretability and data-driven adaptability. For instance, model-based observers such as Kalman filters can provide reliable real-time estimation, while data-driven components compensate for modeling errors, parameter drift, and aging effects. Hybrid frameworks may also incorporate impedance-based features or temperature-dependent corrections to enhance estimation performance [24]. By fusing multiple sources of information, hybrid SOC estimation methods achieve higher accuracy and better generalization compared to standalone techniques, making them particularly suitable for advanced battery management systems.

4.3.2 Depth of Discharge

The DOD of a LIB is the ratio of battery capacity that has been discharged from a completely charged battery to the nominal capacity. The DOD can be expressed as the percentage quantity or in terms of Ah [25]. DOD is a complement to SOC, as SOC increases DOD decreases, and vice versa. For example, if a 100 Ah battery is discharged with a current of 50 A for 40 minutes, the DOD can be calculated as $50 \times 40/60 = 33.33\%$. DOD can be calculated as

$$\text{DOD} = \frac{Q_{\text{released rated capacity}}}{Q_{\text{rated}}}$$

DOD is useful in the efficient utilization of battery capacity preventing deep discharging that accelerates battery aging. DOD also enables predictive maintenance of batteries based on usage patterns and remaining capacity.

4.3.3 State of Health

The SOH of an LIB is defined as the ratio of the battery's capability to retain or hold a certain amount of charge (Q_{max}) to its rated capacity (Q_{rated}) [26].

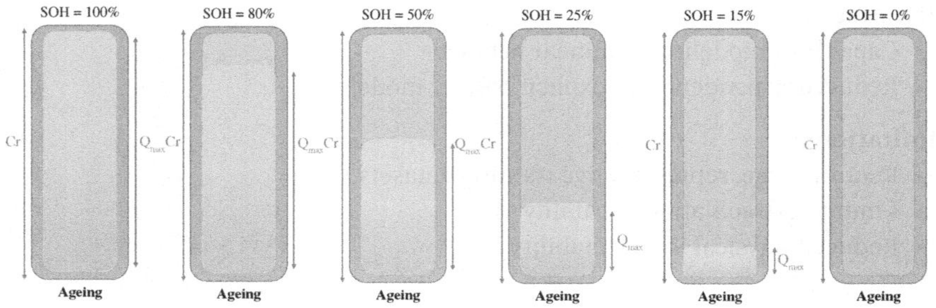

Figure 4.5 State of health with battery aging.

$$\mathrm{SOH} = \frac{Q_{\max}}{Q_{\mathrm{Rated}}}$$

Here, $Q_{\max}$ is the maximum available charge and rated is the rated (nominal) capacity of the battery. The pictorial representation of SOH degradation is as shown in Figure 4.5.

Battery aging occurs due to both electrochemical and mechanical degradation mechanisms. Electrochemical degradation arises from changes in internal chemical composition caused by factors such as high operating temperatures, overcharging, over-discharging, and high current loads. Mechanical degradation is associated with structural changes in electrode materials due to repeated expansion and contraction during charge-discharge cycles. With the increasing demand for fast charging in electric vehicles, batteries are often subjected to high current rates, which can further accelerate degradation processes and reduce battery lifespan. Therefore, accurate estimation of SOH is essential for ensuring safe, reliability, and efficient battery operation.

SOH estimation methods can be broadly classified into experimental methods and model-based methods, based on how degradation information is obtained and processed as shown in Figure 4.6. Experimental methods rely on measurable battery responses and diagnostic testing, whereas model-based methods utilize mathematical models and data-driven algorithms for real-time estimation.

4.3.3.1 Experimental Method

Experimental methods estimate SOH using direct measurements or analysis of battery response characteristics. These methods are widely used in laboratory environments due to their high accuracy and ability to provide insights into degradation mechanisms.

a. Direct assessment methods evaluate SOH using straightforward measurements without complex processing.

   Capacity measurement is one of the most direct and widely used methods for SOH estimation. Using the SOH definition introduced earlier, the maximum

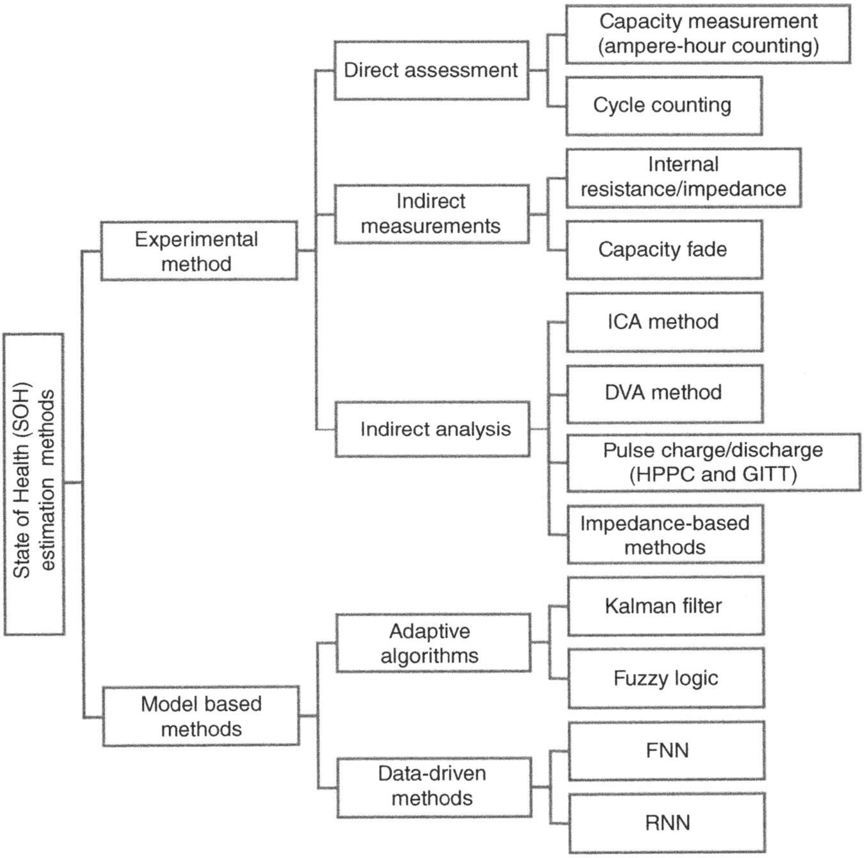

Figure 4.6 SOH estimation methods.

available capacity (Q_{max}) obtained from a full charge-discharge cycle is compared with the rated capacity (Q_{rated}):

$$\text{SOH} = \frac{Q_{\text{max}}}{Q_{\text{Rated}}}$$

This method provides a highly accurate estimate of SOH; however, it requires complete charge-discharge cycles, making it unsuitable for real-time applications in battery management systems.

The cycle counting method estimates the SOH based on the number of charge-discharge cycles. While the approach is simple, it does not account for varying operating conditions such as temperature, depth of discharge, and load profiles, which significantly influence degradation.

b. Indirect measurement methods infer SOH from electrical parameters that evolve with aging.
As the batteries degrade, internal resistance increases due to mechanisms such as SEI layer growth and electrode degradation. This results in higher voltage drop, increased heat generation and reduced power capability. Internal resistance-based methods are fast and non-destructive but may not fully capture the capacity related degradation.

Capacity fade refers to the gradual reduction in the maximum charge a battery can store over time due to aging and degradation mechanisms. In indirect measurement approaches, capacity fade is inferred from operational data such as partial charge-discharge cycles, voltage response, and historical usage patterns, rather than performing full capacity tests. Changes in measurable parameters, including terminal voltage and current profiles, can be used to estimate the loss of available capacity under real operating conditions. Although indirect methods enable practical implementation without interrupting normal operation, their accuracy may be affected by measurement noise, varying load conditions, and incomplete utilization of the full state-of-charge range.

c. Indirect analysis methods extract health-related features from voltage and current response under controlled conditions.
Incremental capacity analysis (ICA) and differential voltage analysis (DVA): These methods analyze derivative curves (dQ/dV and dV/dQ) to identify electrochemical changes within the battery. Variations in peak position, height, and area are associated with specific degradation mechanisms such as loss of active material and Li inventory [27, 28]. However, these methods are sensitive to measuring noise and require filtering techniques.

Pulse-based Methods (HPPC and GITT): These techniques apply controlled current pulses and analyze voltage response to estimate parameters such as internal resistance and diffusion characteristics [29]. While effective for characterizing battery dynamics, they are time-consuming and require specialized equipment.

Impedance-Based Methods (EIS): Electrochemical impedance spectroscopy analyzes battery response over a wide frequency range, enabling separation of different electrochemical processes based on their time constants [21]. EIS provides detailed and non-destructive insight into degradation mechanisms and is highly effective for SOH estimation. However, long testing time and hardware limitations restrict its practical implementation in onboard systems.

4.3.3.2 Model-based Methods

a. Adaptive methods use state observers and filtering techniques to estimate SOH by continuously updating model parameters.
Kalman Filters (KF, EKF, and UKF) are recursive algorithms which estimates internal battery states such as capacity and resistance by combining model predictions with measurable data. They are widely used due to their robustness, real-time capability, and ability to handle nonlinear battery behavior.

Data Driven Methods are increasingly employed to accurately capture the nonlinear behavior of LIBs. These approaches utilize machine learning techniques to establish relationships between measurable variables, such as voltage, current, and temperature, and internal battery states using large datasets. A key advantage of data-driven models is their ability to learn and adapt to changes in battery behavior over time, including aging effects. Their capability to integrate multiple data sources enables a more comprehensive representation of battery dynamics.

In contrast to model-based techniques, which rely on indirect data analysis through mathematical modeling and filtering, data-driven approaches directly learn system behavior from data without requiring explicit physical models. Various machine learning frameworks have been applied for this purpose, including neural networks, Gaussian process regression, ensemble learning, and transfer learning. Additionally, techniques such as fuzzy logic, support vector machines, and decision trees are also widely used for modeling and estimation of battery states. Recent advancements have further explored both recurrent and non-recurrent neural network architectures, particularly for SOC and SOH estimation under dynamic operating conditions.

4.3.4 Remaining Useful Life

While RUL has been discussed in Chapter 2, it is briefly revisited here to maintain continuity and align with the discussion on battery state estimation and prognostics. The prediction of RUL as part of the prognostics and health management (PHM) is extremely important to ensure the reliability and safety of the battery system. The catastrophic failures of the LIB system further emphasized the utmost necessity of RUL prediction [30, 31]. Prediction of RUL in advance ensures operational safety and provides some key information for maintenance and replacement of batteries. The RUL of a battery corresponds to the number of charges and discharge cycles it can go through before reaching its end of life (EOL) [32]. Where EOL generally corresponds to the time when the capacity of the battery is reduced below 70–80% of its nominal capacity [33].

4.3.5 State of Temperature

The temperature is a critical factor that influences the performance, safety, and cycle life of the LIB's [34]. Operation outside the recommended temperature range can accelerate degradation, increase internal resistance, and, in extreme cases, lead to thermal runaway. Therefore, monitoring and controlling battery temperature is essential for safe and reliable operation.

Battery temperature is typically measured using surface-mounted sensors; however, implementing sensors for every cell in a battery pack increases system complexity, cost, and wiring requirements. To address this, sensorless estimation techniques are increasingly used, where the state of temperature (SOT) is estimated using limited sensor data. It is important to note that surface temperature does not accurately represent internal cell temperature, as the core temperature can be significantly higher and exhibits delayed thermal response due to heat transfer effects [35, 36]. These limitations highlight the importance of accurate thermal modeling and effective temperature control strategies, which are addressed through battery thermal management systems (TMSs) in the following section.

4.4 Battery Thermal Management System

A battery thermal management system (BTMS) is an essential subsystem of LIB packs, responsible for maintaining cell temperatures within a safe and optimal operating range. Effective thermal management ensures uniform temperature distribution across cells, enhances performance, prolongs battery life, and prevents safety hazards such as thermal runaway. In high-power applications such as electric vehicles, where batteries are subjected to dynamic load profiles and fast charging, BTMS plays a critical role in ensuring reliable operation.

4.4.1 LIB Thermal Behavior

Thermal management is a critical aspect of LIB systems, particularly in high-power applications such as electrified powertrains. The electrochemical processes within LIBs generate heat through both irreversible and reversible mechanisms, leading to a temperature rise that affects performance, safety, and lifetime [37]. Thermal management in LIBs requires a combination of effective cooling strategies and accurate modeling techniques. While air and liquid cooling dominate current EV applications, emerging technologies such as phase change materials (PCM) and hybrid cooling systems are gaining attention. Lumped thermal models provide a computationally efficient framework for real-time control, while advanced multi-node models improve accuracy. Integrating these models with intelligent control strategies ensures safe, reliability, and high-performance battery operation. The total heat generation within a battery cell can be expressed as:

$$\dot{Q} = I(V_{\mathrm{OCV}} - V_{\mathrm{term}}) - IT\frac{dV_{\mathrm{OCV}}}{dT}$$

where the first term represents irreversible (Joule and polarization) heat and the second term corresponds to reversible (entropic) heat associated with electrochemical reactions [38]. At high current rates, such as those encountered in heavy-duty applications, irreversible heat dominates:

$$\dot{Q}_{\mathrm{irr}} = I^2R(\mathrm{simplified})$$

Effective thermal management is essential to prevent thermal runaway, improve cycle life, and maintain uniform temperature distribution across cells.

4.5 Thermal Management Techniques for EV Battery Packs

TMS in electric vehicles is designed to maintain battery temperature within an optimal operating range (typically 20–40 °C) while minimizing temperature gradients across cells [39].

- Air cooling is the simplest and most cost-effective thermal management approach and is commonly used in low-power or cost-sensitive battery applications [40]. In passive air cooling, heat is dissipated through natural convection and radiation without the use of active airflow, relying on temperature differences between the battery surface and the surrounding environment. While this method is simple and reliable, its heat removal capability is limited. To enhance performance, forced air cooling employs fans or blowers to increase air flow across battery modules, significantly improving convective heat transfer and promoting more uniform temperature distribution through optimized air duct design [41]. Despite its advantages of low cost, lightweight structure, and ease of integration, air cooling suffers from a relatively low heat transfer coefficient (typically 10–50 W/m^2K), resulting in poor temperature uniformity in large battery packs and limited effectiveness under high C-rate operation or fast charging conditions.
- Liquid cooling is widely adopted in modern battery systems due to its superior heat removal capability and effectiveness in maintaining temperature uniformity across cells [42]. In this approach, a coolant is circulated through cold plates or embedded channels in contact with battery modules, transferring

heat to a secondary loop where it is dissipated through a radiator or heat exchanger [43]. Compared to air cooling, liquid cooling offers a significantly higher heat transfer coefficient (typically 200–1000 W/m^2K), making it suitable for high-power applications and fast-charging conditions. However, this method introduces additional system complexity, increased cost and weight, and potential risks associated with coolant leakage and sealing requirements.

- Phase change material (PCM)-based cooling utilizes materials that absorb and store heat through phase transition, typically from solid to liquid. During operation, PCM absorbs excess thermal energy as latent heat, thereby limiting temperature rise and stabilizing battery operation under transient conditions. This passive cooling approach does not require external power and is particularly effective for short-duration high-load scenarios [44]. However, PCM systems generally exhibit low thermal conductivity, which limits heat dissipation rates. As a result, they are often combined with active cooling methods or enhanced with conductive additives such as graphite to improve thermal performance.
- Advanced thermal management techniques are being developed to meet the increasing demands of high-performance and fast-charging battery systems. Refrigerant-based cooling integrates the battery thermal system with the vehicle's HVAC loop, enabling direct heat extraction with high efficiency. Heat pipes and vapor chambers utilize phase-change mechanisms to rapidly distribute heat across battery cells, improving temperature uniformity. Immersion cooling involves submerging battery cells in dielectric fluids, providing highly effective heat removal and uniform cooling [45]. Hybrid cooling systems combine multiple approaches, such as air, liquid, and PCM, to achieve an optimal balance between performance, cost, and system complexity. These advanced methods offer improved thermal regulation but often involve higher design complexity and integration challenges.

4.5.1 Lumped Thermal Modeling of Battery Packs

The lumped thermal model is a widely used approach for analyzing battery temperature due to its simplicity and computational efficiency. In this model, the battery is assumed to have a uniform temperature distribution, neglecting spatial temperature gradients within the cell or pack. The thermal behavior is described using an energy balance equation, where the rate of temperature change is governed by the difference between internally generated heat and heat dissipated to the surroundings [46]. Heat generation is primarily dominated by Joule heating under high current operation, while heat dissipation is modeled using convective heat transfer to the cooling medium. Lumped models are particularly suitable for real-time implementation in battery management systems due to their low computational requirements. However, their accuracy is limited to scenarios involving significant temperature gradients, such as large battery packs or high C-rate operation, where more detailed multi-node or distributed thermal models are required

A widely used approach for real-time thermal analysis is the lumped-parameter thermal model, which assumes spatially uniform temperature distribution.

$$mc_p = \frac{dT}{dt} = \dot{Q} \times \text{gen} - \dot{Q} \times \text{los}$$

where heat generation is approximated as:

$$\dot{Q} \times \text{gen} = I^2 R \times \text{pack}$$

and convective loss is:

$$\dot{Q} \times \text{loss} = hA(T - T \times \text{cool})$$

Combining:

$$\frac{dT}{dt} = \frac{I^2 R_{\text{pack}} - hA(T - T_{\text{cool}})}{mc_p}$$

4.5.1.1 Illustrative Example: Pack Thermal Response

Consider a LIB pack with the following properties:

- Mass $m = 280$ kg
- Specific heat $c_p = 950$ J/kgK
- Effective surface area A = 3.2 m^2
- Coolant temperature $T_{\text{cool}} = 30$ °C
- Initial pack temperature $T_0 = 32$ °C
- Convective heat transfer coefficients:
  - Cooling ON: $h_{\text{ON}} = 220$ W/m^2K
  - Cooling OFF: $h_{\text{OFF}} = 25$ W/m^2K
- Pack internal resistance $R_{\text{pack}} = 0.042$ Ω

  A dynamic current profile applied is as follows:
  - 0–500 s: I = 180 A
  - 500–1000 s: I = 120 A
  - 1000–1500 s: I = 0 A (rest)

The cooling system is assumed to remain OFF during the first 500s and then switch ON for the remainder of the cycle. The objective is to estimate the heat generation, steady-state temperature levels, and approximate transient temperature evolution.

Step 1: Thermal capacitance of the pack

The effective thermal capacitance is

$$C_{\text{th}} = mc_{\text{p}} = 280 \times 950 = 266{,}000\ \text{J/K}$$

This means that 266,000 J of heat is required to raise the pack temperature by 1 °C.

Step 2: Heat generation during each interval

The Joule heat generation is approximated by

$$\dot{Q} \times \text{gen} = I^2 R \times \text{pack}$$

For the first interval $(I = 180\ \text{A})$,

$$\dot{Q}_{\text{gen},1} = 180^2 \times 0.042 = 1360.8\ \text{W}$$

For the second interval (I = 120 A),

$$\dot{Q}_{\text{gen},2} = 120^2 \times 0.042 = 604.8\ \text{W}$$

For the resting period,

$$\dot{Q}_{\text{gen},3} = 0^2 \times 0.042 = 0\ \text{W}$$

Step 3: Steady-state temperature for each operating mode

The steady-state temperature is obtained from

$$T_{\text{ss}} = T_{\text{cool}} + \frac{\dot{Q}_{\text{gen}}}{hA}$$

Case 1:

First interval, cooling OFF

$$hA = h_{\text{OFF}}A = 25 \times 3.2 = 80\ \text{W/K}$$

Thus,

$$T_{\text{ss},2} = 30 + \frac{1360.8}{80} = 30 + 17.01 = 47.01\ \text{°C}$$

Case 2:

Second interval, cooling ON

$$hA = h_{\text{ON}}A = 220 \times 3.2 = 704\ \text{W/K}$$

Hence,

$$T_{\text{ss},2} = 30 + \frac{604.8}{704} = 30 + 0.86 = 30.86\ \text{°C}$$

Case 3:

Rest period, cooling ON

Since $\dot{Q}_{\text{gen}} = 0$

$$T_{\text{ss},3} = T_{\text{cool}} = 30\ °\text{C}$$

These values clearly show that the pack tends toward a much higher equilibrium temperature when cooling is OFF, whereas active cooling dramatically suppresses the thermal rise.

Step 4: Thermal time constants

The thermal time constant is

$$\tau = \frac{mc_p}{hA}$$

Cooling OFF:

$$\tau_{\text{OFF}} = \frac{266{,}000}{80} = 3325\ \text{s} = 55.4\ \text{min}$$

Cooling ON:

$$\tau_{\text{ON}} = \frac{266000}{704} = 377.8\ \text{s} = 6.3\ \text{min}$$

This indicates that the battery temperature changes very slowly when cooling is OFF but responds much faster once active cooling is engaged.

Step 5: Approximate transient temperature response

For a first order lumped thermal model with constant current in each interval, the temperature evolution can be approximated as

$$\boldsymbol{T(t) = T_{\text{ss}} + (T_{\text{init}} - T_{\text{ss}})e^{\frac{-t}{\tau}}}$$

Interval 1: 0-500 s, 180 A, cooling OFF: Initial temperature is $T_0 = 32\ °\text{C}$, with $T_{\text{ss},1} = 47.01\ °\text{C}$ and $\tau_{\text{OFF}} = 3325$ s. Therefore, at t = 500 s,

$$T(500) = 47.01 + (32 - 47.01)e^{\frac{-500}{3325}}$$
$$T(500) = 47.01 - (15.01)e^{-0.1504}$$
$$T(500) = 47.01 + (15.01) * 0.8604 = 34.10\ °\text{C}$$

Interval 2: 500–1000 s, 120 A, cooling ON Now the initial temperature is 34.10 °C, with

$T_{\text{ss},2} = 30.86\ °\text{C}$ and $\tau_{\text{ON}} = 377.8$ s. After another 500 s,

$$T(1000) = 30.86 + (34.10 - 30.86)e^{\frac{-500}{377.8}}$$
$$T(1000) = 30.86 + (3.24) \times e^{-1.323}$$
$$T(1000) = 30.86 + (3.24) \times 0.2664 = 31.72\ °\text{C}$$

Interval 3: 1000–1500 s, rest, cooling ON: Now the initial temperature is 31.72 °C, and the steady-state temperature is 30 °C. Thus,

$$T(1500) = 30 + (31.72 - 30)e^{\frac{-500}{377.8}}$$
$$T(1500) = 30 + 1.72 \times e^{-1.323}$$
$$T(1500) = 30 + 1.72 \times 0.2664 = 30.46\ °\mathrm{C}$$

The example worked shows several important physical insights. During the first interval, the battery experiences substantial Joule heating, but because the thermal time constant under cooling-OFF conditions is large, the pack temperature rises gradually rather than instantaneously. Once active cooling is enabled, the effective heat rejection increases significantly, the thermal time constantly decreases, and the pack temperature quickly moves toward a lower equilibrium value. During the final rest period, the absence of internal heat generation allows the pack to cool back toward the coolant temperature. From a battery management perspective, this example illustrates that thermal response depends not only on current magnitude, but also on the interplay among pack resistance, thermal capacitance, and cooling effectiveness. It also highlights why proactive cooling activation is often preferred in EV battery packs, particularly during sustained medium- to high-current operation.

4.6 Battery Management Systems Design

In an electric vehicle, the battery pack provides electrical power to the traction inverter, electric motor, and auxiliary vehicle systems [47]. While driving, the battery delivers high current to support acceleration and sustained cruising, while during regenerative braking it absorbs energy recovered from the motor operating as a generator. The battery pack must therefore operate reliably over a wide range of power demands, temperatures, and driving conditions. Its integration with the vehicle powertrain, thermal management system, and battery management system ensures that energy delivery is efficient, safe, and responsive to real-time operating requirements, while also supporting long-term durability and consistent vehicle performance. To monitor individual cell performance in the battery pack during its operation, an onboard battery management system (BMS) is required.

In many respects, the automotive BMS performs a role analogous to that of an engine management system in internal combustion engine vehicles. It is responsible for ensuring that the battery pack operates safely, efficiently, and reliably under all driving conditions. To achieve this, the BMS continuously monitors key battery parameters such as individual cell voltages, pack current, and temperatures distributed across the battery system. Beyond basic monitoring, the BMS plays a central role in battery control and protection by estimating critical internal states that cannot be measured directly. These include the SOC, SOP, SOH, and RUL. Accurate estimation of these states enables effective energy management, power delivery, and predictive maintenance throughout the lifetime of an EV [48, 49]. The BMS also enforces operational limits specified by battery manufacturers, preventing unsafe conditions such as overcharging, over-discharging, excessive current draw, and thermal excursions by regulating voltage, current, and temperature within predefined safe boundaries. To maintain safety and performance, the BMS must independently monitor and manage individual cells while coordinating pack-level operation. In addition to measurement and state estimation functions, the BMS generates control actions such as cell balancing commands, charge and discharge current limits, and fault detection signals. Through these combined monitoring, estimation, and control functions, the BMS ensures optimal utilization of the battery pack while supporting long-term durability and safe operation under the highly dynamic conditions of electric vehicle use. The critical functionalities are summarized in Table 4.1.

Automotive BMS are commonly classified according to their architectural configuration and degree of system distribution. The choice of BMS architecture is strongly influenced by battery pack size, voltage level, scalability requirements, and safety considerations. Broadly, automotive BMS architectures can be categorized into centralized, modular and scalable, wireless, and cloud-based systems, each offering distinct trade-offs in complexity, flexibility, and functionality.

In a centralized BMS architecture, all sensing, monitoring, and control functions are implemented on a single electronic control unit. Individual cell voltages, temperatures, and pack current are directly routed to this central controller, where data acquisition and decision-making are performed by an onboard processor. This architecture offers a compact and cost-efficient solution, suited for small battery packs with a few numbers of cells. The integration of all functionalities onto a single board simplifies system design and reduces hardware cost. However, as battery packs scale in size, centralized BMS architecture become increasingly impractical. The large number of direct electrical connections required leads to extensive wiring harnesses, increased installation effort, and higher susceptibility to wiring-related faults. Consequently, centralized BMS are generally unsuitable for high-voltage, large-capacity battery packs used in modern electric vehicles.

Modular and scalable BMS architectures distribute monitoring and control tasks across multiple interconnected subsystems. In this configuration, a master controller coordinates overall battery operation, while multiple subordinate modules are assigned to individual battery modules. Each subordinate module locally monitors cell-level voltages and temperatures, and in some cases currents, within its assigned module. Communication between the master and subordinate units is typically implemented using structured topologies such as daisy-chain

Table 4.1 Critical functionalities of battery management system.

BMS function	Primary objective	Key measured/ estimated parameters	Typical implementation approaches
Voltage, current, and temperature monitoring	Ensure safe operation and provide inputs for control and estimation algorithms	Cell voltage, pack voltage, current, cell and pack temperature	Voltage sensing ICs, shunt or Hall-effect current sensors, thermocouples or RTDs
Charge and Discharge Control	Maintain operation within manufacturer-defined safe limits during charging and discharging	Voltage limits, current limits, temperature thresholds	CC-CV charge control, over-/ under-voltage protection, current limiting
State estimation	Provide real-time insight into battery condition and usable energy	SOC, SOH, SOP, SOE, RUL	Coulomb counting, model-based observers (Kalman filters), data-driven approaches
Cell voltage and SOC balancing	Reduce cell imbalance and maximize usable pack capacity	Individual cell voltages, SOC deviation	Passive resistive balancing, active energy redistribution circuits
Thermal management system control	Maintain battery temperature within optimal range for safety and longevity	Cell and pack temperatures, thermal gradients	Active cooling/ heating control, feedback to liquid or air-based TMS
Isolation detection	Ensure electrical safety in high-voltage battery systems	Isolation resistance between terminals and chassis	Continuous insulation monitoring, fault-triggered protection
Communication and data logging	Enable coordination with vehicle systems and long-term battery analytics	SOC, SOP, SOH, temperature, current history	CAN/CAN-FD communication, onboard logging, IoT and cloud-based analytics

or multi-drop networks, enabling straightforward scalability as battery pack size increases. The use of dedicated analog front-end devices allows accurate and synchronized measurement across large cell stacks while significantly reducing wiring complexity. By placing measurement electronics closer to the battery modules, modular BMS architectures improve signal integrity, simplify maintenance, and enhance fault isolation. These characteristics make modular and scalable BMS the preferred architecture for contemporary electric vehicles with large and complex battery packs.

Wireless and cloud-enabled battery management systems represent an evolution of conventional BMS architectures aimed at improving scalability, flexibility, and data utilization in large battery packs. Wireless BMS replace extensive wired communication networks with wireless links between distributed sensing nodes and central controller, significantly reducing wiring complexity, system weight, and connector-related reliability issues. This approach enables greater design flexibility and simplifies pack integration, particularly in high-capacity battery systems. Cloud-enabled BMS extend this concept by transmitting battery data beyond the vehicle to remote computing platforms, where advanced diagnostics, prognostics, and optimization algorithms can be executed. By leveraging cloud computing resources, these systems support computationally intensive functions such as long-term health monitoring, digital-twin-based state prediction, and fleet-level analytics that are impractical for onboard controllers alone. Together, wireless and cloud-enabled BMS architectures enable scalable, data-driven battery management while introducing new challenges related to communication reliability, cybersecurity, latency, and large-scale validation.

References

1 A. Alem, T. Kalogiannis, J. Van Mierlo, and M. Berecibar, "A comprehensive review of stationary energy storage devices for large scale renewable energy sources grid integration," vol. 159, 2022.

2 M. Kandhasamy, B. K. Duvaragan, and S. Kamaraj, "An overview on classification of energy storage systems," in *Mater. Boost. Energy Storage.* Vol. 2 *Adv. Sustain. Energy Technol. Part 1—An Overv. Classif. Energy Storage Syst.*, vol. 2, pp. 1–25, 2024.

3 F. M. N. Uddin, M. G. Rasul, A. S. M. Sayem, and N. K. Mandal, "Design and optimization of lithium-ion battery as an efficient energy storage device for electric vehicles : A comprehensive review," J. *Energy Storage*, vol. 71, 2023.

4 L. Zhou et al., "State estimation models of lithium-ion batteries for battery management system: Status, challenges, and future trends," *Batteries*, vol. 9, no. 2, 2023.

5 S. Jafari and Y. C. Byun, "Prediction of the battery state using the digital twin framework based on the battery management system," *IEEE Access*, vol. 10, pp. 124685–124696, 2022.

6 D. Bernardi and J. Electrochem, "A general energy balance for battery systems," *J. Electrochem. Soc.*, 1985.

7 S. Cho, J. Yoon, Y. S. Yun, and H. J. Jin, "Improved cyclability of lithium-ion batteries using pyroprotein-assisted silicon anodes," *ACS Appl. Energy Mater.*, vol. 5, no. 12, pp. 15538–15547, 2022.

8 J. Wang et al., "Improving cyclability of Li metal batteries at elevated temperatures and its origin revealed by cryo-electron microscopy," *Nat. Energy*, vol. 4, no. 8, pp. 664–670, 2019.

9 D. Versaci et al., "Tailoring cathode materials: A comprehensive study on LNMO/LFP blending for next generation lithium-ion batteries," *J. Power Sources*, vol. 613, p. 234955, 2024.

10 F. Baskoro, S. U. Sharma, A. L. Lubis, and H. J. Yen, "Recent advances in p-type polymeric electrode materials towards high-voltage 4.0 V-class organic lithium-ion batteries," *J. Mater. Chem. A*, vol. 13, no. 3, pp. 1552–1589, 2024.

11 S. Sokhanvaran, S.-K. Lee, G. Lambotte, and A. Allanore, "Electrochemistry of molten sulfides: Copper extraction from BaS-Cu 2 S," *J. Electrochem. Soc.*, vol. 163, no. 3, pp. D115–D120, 2016.

12 H. S. Ramadan, M. Becherif, and F. Claude, "Extended kalman filter for accurate state of charge estimation of lithium-based batteries: a comparative analysis," *Int. J. Hydrogen Energy*, 2017.

13 M. Yue, S. Jemei, R. Gouriveau, and N. Zerhouni, "Review on health-conscious energy management strategies for fuel cell hybrid electric vehicles: Degradation models and strategies," *Int. J. Hydrog. Energy*, 2019.

14 Q. Liao et al., "Performance assessment and classification of retired lithium ion battery from electric vehicles for energy storage," *Int. J. Hydrogen Energy*, 2017.

15 J. Zhang and J. Lee, "A review on prognostics and health monitoring of Li-ion battery," *J. Power Sources*, 2011.

16 X. Wu, X. Li, and J. Du, "State of charge estimation of lithium-ion batteries over wide temperature range using unscented Kalman filter," *IEEE Access*, 2018.

17 Y. Wang et al., "A comprehensive review of battery modeling and state estimation approaches for advanced battery management systems," *Renew. Sustain. Energy Rev.*, vol. 131, 2020.

18 J. Meng et al., "Overview of Lithium-Ion battery modeling methods for state-of-charge estimation in electrical vehicles," *Appl. Sci.*, vol. 8, no. 5, 2018.

19 K. Liu et al., "Electrochemical modeling and parameterization towards control-oriented management of lithium-ion batteries," *Control Eng. Pract.*, vol. 124, 2022.

20 J. Zhou, B. Xing, and C. Wang, "A review of lithium ion batteries electrochemical models for electric vehicles," *E3S Web Conf.*, vol. 185, pp. 10–12, 2020.

21 R. Xiong, L. Li, and J. Tian, "Towards a smarter battery management system: A critical review on battery state of health monitoring methods," *J. Power Sources*, vol. 405, pp. 18–29, 2018.

22 M. Chen, G. Ma, W. Liu, N. Zeng, and X. Luo, "An overview of data-driven battery health estimation technology for battery management system," *Neurocomputing*, vol. 532, pp. 152–169, 2023.

23 H. M. Fahmy, et al., "Hybrid State of Charge Estimation of Lithium-Ion Battery Using the Coulomb Counting Method and an Adaptive Unscented Kalman Filter," *Energies*, 2023.

24 Y. Yu, and J. Liang, "Hybrid state-of-charge estimation for lithium-ion batteries: A structured survey," *Syst. Sci. Control Eng.*, vol. 14, no. 1, 2026.

25 K. Das and R. Kumar, "Electric vehicle battery capacity degradation and health estimation using machine-learning techniques: A review," *Clean Energy*, vol. 7, no. 6, pp. 1268–1281, 2023.

26 X. Shu et al., "State of health prediction of lithium-ion batteries based on machine learning: Advances and perspectives," *iScience*, vol. 24, no. 11, 2021.

27 A. Fly, B. Wimarshana, I. Bin-Mat-Arishad, and M. Sarmiento-Carnevali, "Temperature dependency of diagnostic methods in lithium-ion batteries," *J. Energy Storage*, vol. 52, 2022.

28 L. Spitthoff, P. J. S. Vie, M. S. Wahl, J. Wind, and O. S. Burheim, "Incremental capacity analysis (dQ/dV) as a tool for analysing the effect of ambient temperature and mechanical clamping on degradation," *J. Electroanal. Chem.*, vol. 944, 2023.

29 T. Białoń, R. Niestrój, W. Skarka, and W. Korski, "HPPC test methodology using lfp battery cell identification tests as an example," *Energies*, vol. 16, no. 17, 2023.

30 Q. Wang, et al., "Thermal runaway caused fire and explosion of lithium ion battery," *J. Power Sources*, 2012.

31 H. Tian, P. Qin, K. Li, and Z. Zhao, "A review of the state of health for lithium-ion batteries: Research status and suggestions," *J. Clean. Prod.*, 2020.

32 X. Hu, L. Xu, X. Lin, and M. Pecht, "Battery lifetime prognostics," *Joule*, vol. 4, no. 2, pp. 310–346, 2020.

33 Y. H. Lui et al., "Physics-based prognostics of implantable-grade lithium-ion battery for remaining useful life prediction," *J. Power Sources*, vol. 485, p. 229327, 2021.

34 F. Feng et al., "Co-estimation of lithium-ion battery state of charge and state of temperature based on a hybrid electrochemical-thermal-neural-network model," *J. Power Sources*, vol. 455, p. 227935, 2020.

35 R. R. Richardson, P. T. Ireland, and D. A. Howey, "Battery internal temperature estimation by combined impedance and surface temperature measurement," *J. Power Sources*, vol. 265, pp. 254–261, 2014.

36 S. Surya, A. Samanta, V. Marcis, and S. Williamson, “Hybrid electrical circuit model and deep learning-based core temperature estimation of lithium-ion battery cell,” *IEEE Trans. Transp. Electrif.*, vol. 8, no. 3, pp. 3816–3824, 2022.

37 H. Chun, H. Choi, Y. Jun, and H. Lee, “Comprehensive study on thermal characteristics of lithium-ion battery with entropic heat,” *Int. J. Energy Res.*, 2024.

38 C. Liu et al., “Review—understanding thermal runaway in lithium-ion batteries: trigger, mechanism, and early warning strategies,” *J. Electrochem. Soc.*, 2024.

39 Q. Wang, B. Jiang, B. Li, and Y. Yan, “A critical review of thermal management models and solutions of lithium-ion batteries for the development of pure electric vehicles,” vol. 64, pp. 106–128, 2016.

40 X. Zhang, Z. Li, L. Luo, Y. Fan, and Z. Du, “A review on thermal management of lithium-ion batteries for electric vehicles,” vol. 238, 2022.

41 A. A. H. Akinlabi and D. Solyali, “Configuration, design, and optimization of air-cooled battery thermal management system for electric vehicles : A review,” vol. 125, 2020.

42 C. Wu, Y. Sun, H. Tang, S. Zhang, W. Yuan, and L. Zhu, “A review on the liquid cooling thermal management system of lithium-ion batteries,” vol. 375, 2024.

43 C. Roe et al., “Immersion cooling for lithium-ion batteries – A review,” *J. Power Sources*, vol. 525, p. 231094, 2022.

44 A. G. Mohammed, K. E. Elfeky, and Q. Wang, “Recent advancement and enhanced battery performance using phase change materials based hybrid battery thermal management for electric vehicles,” vol. 154, 2022.

45 A. Gharehghani et al., “Progress in battery thermal management systems technologies for electric vehicles,” vol. 202, 2024.

46 X. Lin et al., “Online parameterization of lumped thermal dynamics in cylindrical lithium ion batteries for core temperature estimation and health monitoring,” vol. 21, pp. 1745–1755, 2013.

47 A. K. M. A. Habib, et al., “A review: Energy storage system and balancing circuits for electric vehicle application,” *IET Power Electron.*, vol. 14, no. 1, pp. 1–13, 2021.

48 A. M. K. A. Opitz, P. Badami, L. Shen, K. Vignarooban, “Can Li-ion batteries be the panacea for automotive applications?,” vol. 68, pp. 685–692, 2017.

49 S. Ahmed et al., “Cost of automotive lithium-ion batteries operating at high upper cutoff voltages,” *J. Power Sources*, vol. 403, pp. 56–65, 2018.

5

Battery Charging Systems

5.1 Introduction

The widespread adoption of transportation electrification and battery energy storage systems (BESSs) has increased the importance of safe and efficient charging of lithium-ion batteries (LIBs) [1]. Although these two applications differ in operating profiles and system requirements, they share a common electrochemical foundation and charging restrictions. Charging plays a crucial role in both the applications to determine the lifetime, efficiency, safety, and performance of LIBs [2, 3]. As a result, the design of the charging system must carefully consider the battery characteristics, power-electronic interfaces, thermal safety, and grid level constraints. Electric vehicle (EV) batteries are often charged under time-critical conditions, where minimizing the charging duration is essential for user acceptance and vehicle operation. However, BESSs are frequently charged according to grid-based schedules, where charging may be optimized to maximize efficiency, reduce operating cost, or to accommodate grid demand [4]. Nevertheless, in both cases, the charging of the batteries must be performed considering the defined voltage, current, and temperature limits to prevent unsafe operating conditions and accelerated battery degradation.

In EV applications, the charging systems should accommodate a wide range of use cases, from slow residential charging to high-power public fast charging. The charging strategy influences long-term capacity retention, safe thermal behavior, and vehicle downtime. In contrast, the grid storage charging systems are optimized for grid services such as peak shaving, frequency regulation, renewable energy smoothing, and backup power. The applications involve scheduled or repetitive charging and discharging, making battery degradation-aware charging essential. Although charging strategies for EV and BESSs are different, many charging methods are equally applicable to both the applications considering the system constraints [5].

The EV charging is constrained by battery electrochemical limits, thermal safety boundaries, power electronics, and grid constraints. Charging batteries without proper regulation will lead to excessive heat generation, uneven current distribution among cells, and may lead to unwanted side reactions. Over time, these effects contribute to the fading of capacity, increased internal resistance, and reduced usable energy [6]. In extreme conditions, improper charging may lead to safety risk of thermal runaway. The design of effective charging systems, therefore, requires a multidisciplinary approach that integrates battery electrochemistry, power electronics, thermal management, and system-level control. Power-electronic converters must regulate current and voltage with high accuracy while operating efficiently across a wide range of power levels. Thermal management systems must remove excess heat generated during charging and maintain the battery within an acceptable temperature window. At the same time, control algorithms implemented within the battery management system (BMS) should monitor key variables such as cell voltage, current, and temperature, and adjust charging behavior accordingly [7].

At the cell and pack levels, battery charging is commonly controlled using predefined current and voltage profiles implemented through power-electronic converters and supervised by a BMS. The most widely adopted approach is constant current-constant voltage (CC-CV) charging, which provides a simple and reliable means of achieving full charge while respecting voltage limits [8]. However, alternative strategies such as multistage constant current (MCC) charging, pulse charging, and temperature-regulated charging have been proposed to improve charging speed, reduce thermal stress, and mitigate aging [9, 10]. These methods are of growing interest in both EV and BESS applications, particularly as systems move toward higher energy densities and more aggressive operating conditions. Beyond charging algorithms, modern battery charging systems must also comply with hardware and communication standards that ensure interoperability, safety, and coordinated power delivery. For EVs, international charging standards define connector geometries, electrical ratings, and digital communication protocols between the vehicle and the charging equipment. Similarly, grid-connected BESS relies on standardized power-conversion interfaces and control architectures to integrate safely with utility networks. In both cases, charging power levels ranging from low-power AC charging to high-power DC fast charging are considered for reducing the charging times.

Battery charging systems should be seen as integrated systems that function at the nexus of electrochemistry, power electronics, communication, and grid interaction rather than just as power delivery methods. Battery properties, charging methodologies, infrastructural capabilities, and external limitations imposed by the electrical grid all have an impact on the feasible charging performance, safety, and lifetime. Because of this, charging techniques cannot be assessed separately; rather, they must be comprehended within a system-level framework that takes operational goals, control coordination, and hardware constraints into account. This viewpoint is equally applicable to stationary BESSs and electric automobiles, which have different use cases and scales but similar charging processes.

This chapter provides a structured and systematic overview of battery charging methods, charging standards, and charging power levels relevant to both EV and BESS applications. The discussion begins with fundamental concepts and definitions required to understand battery charging behavior. Commonly used charging methods are then presented, with emphasis on their operating principles, advantages, and limitations. Major international charging standards and charging levels are subsequently reviewed. Finally, the chapter concludes with solved numerical examples that illustrate practical calculations related to charging time, temperature rise, charging rate selection, and charger sizing. Together, these topics provide a foundation for understanding and designing battery charging systems for transportation and grid energy storage applications.

5.2 Fundamentals of Battery Charging

State of charge (SOC), battery capacity, C-rate, charging current, voltage limits, charge cut-off, charging efficiency, and losses are some of the fundamental concepts required for understanding battery charging behavior in both EV and BESS applications [11]. These fundamentals are the basis for charging methods, standards, and design calculations.

The SOC defines the amount of charge that is remaining in the battery relative to its usable capacity. It is typically expressed in percentage, where 100% corresponds to a fully charged state and 0% corresponds to fully discharged state [12].

$$\mathrm{SOC}(t) = \frac{Q(t)}{Q_{\mathrm{usable}}} \times 100\%$$

where $Q(t)$ is the charge stored in the battery at time t and Q_{usable} is the usable capacity of the battery.

For vehicle applications, SOC is one of the critical parameters for range estimation and for user information to understand the charge level of the battery. In BESS applications, it is used to schedule charging and discharging events for grid services. Accurate SOC estimation is essential for safe charging, because overcharging beyond the allowable SOC and voltage range will lead to excessive voltage rise and fast degradation. C-rate is a normalized measure of charging or discharging current relative to the battery capacity. It provides a convenient way to compare the charging conditions across batteries of different sizes.

$$C - \text{rate} = \frac{I}{Q_{\text{nom}}}$$

where I represents the charging current and Q_{nom} represents the nominal capacity.

EV fast charging often involves C-rates greater than 1C, especially during DC fast charging, whereas BESS charging typically operates at lower C-rates to prioritize efficiency and longevity. High C-rate charging increases internal losses and heat generation, which must be carefully managed to avoid thermal and electrochemical stress. The charging and discharging speed of BESS is determined by C-rate. At a 1C rate, the battery can be charged in one hour. For a 10 MWh BESS operating at 1C, it can be recharged entirely in one hour if supplied with 10 MW of power. This high rate is good for applications requiring rapid energy availability, such as emergency support and grid stabilization. With 0.5C rate, the battery can be charged with 5 MW of power for two hours. The moderate 0.5C rate is suitable for applications like load leveling and peak shaving. The slower rate, such as 0.25C, is beneficial for long-duration energy storage applications, such as storing excess renewable energy generated during off-peak times.

Inorder to ensure reliability and safety, LIBs should be charged under strict voltage limits. Most lithium chemistries have maximum voltage limits, typically ranging from 4.1 to 4.3 V per cell to which they can be safely charged. Charging cells above the maximum voltage will accelerate lithium plating [13]. If the battery is charged to less than the maximum voltage, cycle life may be increased but useful energy per cycle willbe reduced. The charging systems on both the cell and pack levels, therefore, impose voltage cut-off limits. In BESS, such constraints are often interfaced with the inverter and grid controllers, while in EVs, they are realized by a BMS. The BMS communicates with onboard or offboard chargers in EVs to guarantee safe operation in a variety of user scenarios. To satisfy grid requirements while maintaining battery health in BESS applications, the BMS frequently works in tandem with higher-level energy management systems.

5.3 Charging Strategies

5.3.1 Constant Current-constant Voltage

The constant current and constant voltage method of charging is the most widely used strategy for LIBs in both EVs and BESS applications. This is due to the simplicity, robustness, and ability to meet the voltage and current constraints set by the battery manufacturers. The CC-CV charging consists of two sequential phases, a constant current (CC) phase followed by a constant voltage (CV)

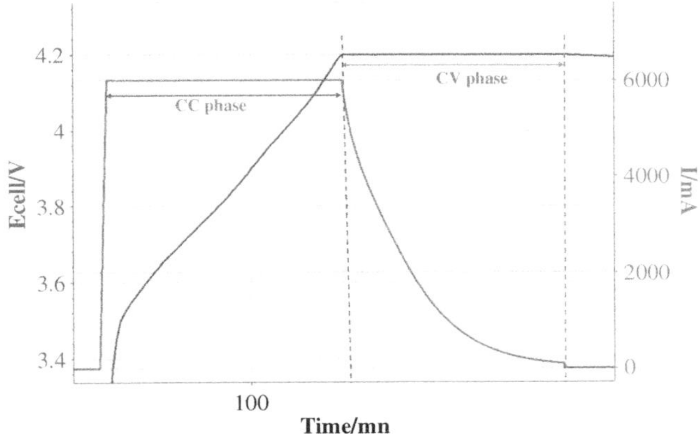

Figure 5.1 Constant-current and constant-voltage curves for a lithium-ion battery.

phase as shown in Figure 5.1. During the CC phase, the battery is charged with a constant current level, and the terminal voltage gradually increases. The charging current is determined by the C-rate of the battery, thermal limits, and charging time or efficiency at the system level. The battery will often be charged to 80% SOC during this phase, which usually accounts for a significant amount of the overall charge transferred. This stage is crucial for EV applications since it dictates how quickly usable driving range can be restored. In BESS applications, the CC phase enables predictable power absorption from the grid or renewable sources.

When the battery terminal voltage reaches the preset maximum voltage limit, V_{max}, the charging switches to the CV phase. During the CV phase, the battery voltage is held constant at V_{max}, and while the charging current gradually decreases as the battery approaches full SOC. The charging process is typically terminated when the current falls below a specified threshold, such as C/10 or C/20. This CV phase ensures that the battery reaches full SOC without exceeding voltage limits that could trigger harmful side reactions. From an electrochemical aspect, during the CC phase, the lithium ions are intercalated effectively into the electrodes because polarization effects are low polarization moderate SOCs. As SOC increases, the lithium-ion concentration gradients within the electrodes become more pronounced leading to high polarization and electrochemical overpotential. The natural reduction of the current during the CV phase prevents additional polarization and lowering the possibility of lithium plating near the end of charge.

Note: Polarization here refers to the deviation of terminal voltage from the equilibrium voltage of the battery due to internal resistance and also due to the kinetic and mass transport limitations.

In EV applications, both onboard chargers and offboard fast chargers perform CC-CV charging. High currents are provided during the CC phase of DC fast charging to reduce charging time, whereas the CV phase frequently becomes the primary contributor to the overall charging length. One well-known drawback of CC-CV charging in fast-charging situations, particularly for high-energy battery packs, is this longer CV tail. Because large currents during the CC phase can produce a significant amount of heat, thermal management is essential during EV fast charging. In order to modify the charging current or stop charging if safety limits are reached, the BMS continuously checks cell voltages and temperatures. Bidirectional power converters and CC-CV charging are frequently utilized in grid-connected BESS. While the CV phase guarantees complete charge without overstressing the cells, the CC phase enables the BESS to absorb a specified amount of electricity from the grid or renewable energy sources. BESS usually have less aggressive charging schedules and lower C-rates than EVs. As a result, charging efficiency is frequently higher and the CV phase may be less noticeable. However, for controlling energy flow in stationary storage systems, CC-CV charging offers a dependable and consistent framework.

The CC-CV charging provides a clear and proven implementation that closely complies with voltage and current constraints given by the manufacturer, making it easy to monitor using conventional BMS architectures. Its organized two-stage management allows for safe and dependable charging in a variety of applications, enabling the same basic method to be used for both stationary BESS and EVs. However, the extended constant-voltage phase, which increases the charging time, is an intrinsic drawback of CC-CV charging. Furthermore, the fixed nature of the charging profile restricts flexibility to changes in temperature, aging, and operating circumstances, and high current operation during the constant-current phase can result in enhanced thermal stress, especially at high C-rates. Engineers must thus carefully choose the constant-current level when implementing CC-CV charging based on permitted C-rate and thermal constraints, specify the maximum charging voltage based on the particular cell chemistry, and set a suitable current cut-off threshold to end the constant-voltage phase.

5.3.2 Multistage Constant Current

The MCC charging strategy is an adaptation of the traditional CC-CV method. Here, the overall charging process is divided into multiple constant-current segments with progressively decreasing current levels as shown in Figure 5.2, unlike the conventional CC-CV approach, which applies a single constant current followed by a constant-voltage phase [10]. MCC charging regulates the charging current in discrete stages while continuously respecting the maximum allowable

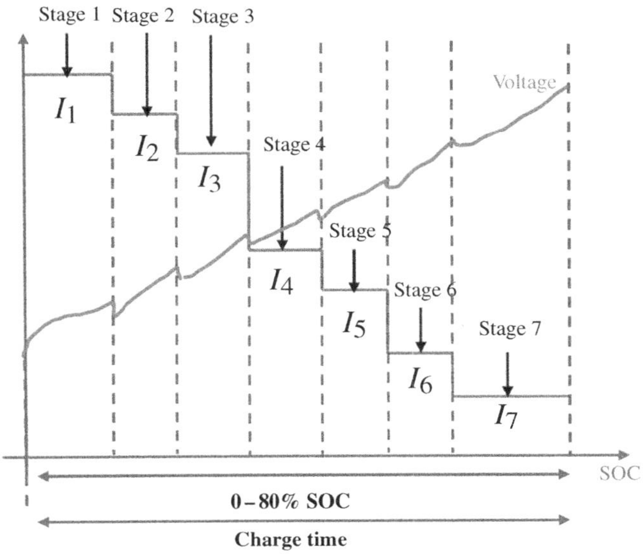

Figure 5.2 The charging process of multistage constant-current charging protocol.

voltage. This structured current reduction enables improved control of electrochemical polarization, temperature rise, and degradation mechanisms.

The primary rationale for MCC charging is based on the observation that LIBs receive high charge at low SOC and increase internal resistance as SOC increases. During the initial charging stage, when polarization effects are minor, a relatively large current can be safely applied. As the battery enters the mid- and high-SOC areas, the charging current is lowered in predetermined increments to limit overpotential, prevent lithium plating, and reduce heat generation [14]. From an electrochemical standpoint, MCC charging reduces the abrupt increase in concentration and kinetic overpotentials that are common during high-SOC charging. By lowering the applied current before the battery reaches voltage saturation, MCC charging reduces the conditions that encourage lithium plating on the graphite anode and excessive solid-electrolyte interphase (SEI) development. Consequently, MCC charging has been demonstrated to lower temperature rise, enhance charge usage, and lengthen cycle life when compared to normal CC-CV charging, particularly under fast-charging situations [9, 15].

In EV applications, MCC charging is suitable for fast-charging scenarios where minimizing charge time without compromising battery safety is the primary goal. An MCC charging strategy must be carefully designed around three critical parameters: the number of charging stages, the current (or C-rate) assigned to each stage, and the transition criterion. Common transition criteria include elapsed time, SOC thresholds, voltage thresholds, and voltage cut-off limits. Voltage-based transitions are particularly popular because they are simple and robust, relying on directly quantifiable quantities and are easily enforced by battery management systems. SOC-based transitions, on the other hand, can provide more accurate control when SOC estimates are reliable [16]. Although MCC charging requires more control complexity than CC-CV charging, it is still substantially easier to design than fully adaptive or model-predictive charging schemes. As a result, MCC charging is an effective intermediate solution that balances charging speed, battery health, and implementation feasibility for both EV and grid-connected BESS applications.

5.3.3 Pulse Charging

Balancing LIB charging duration and cycle life is a challenge. Conventional fast charging methods will affect the cycle life of the battery which increases the voltage polarization due to internal resistance build up. Pulse charging represents an effective charging approach to mitigate the buildup of polarization voltage as shown in Figure 5.3. Pulse charging refers to a method of charging where the battery receives current in a series of pulses followed by intervals of rest or reverse current, rather than a steady flow of current. In contrast to methods that rely on a constant current, pulse charging takes advantage of the varying electrochemical and thermal responses of LIBs over time to enhance their ability to accept charges, especially in challenging conditions like low temperatures or during

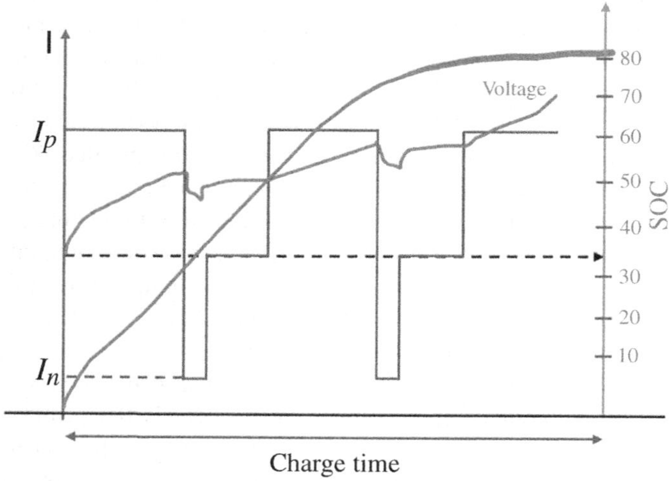

Figure 5.3 Pulse charging protocol.

rapid charging [17]. The applied current switches between charging and relaxation phases during pulse charging. While the pulse duty cycle controls the average charging rate, the instantaneous current during the pulse might be much larger than the corresponding continuous charging current. By allowing concentration gradients within the electrodes and electrolyte to relax during the rest interval, polarization and internal resistance accumulation are decreased. Therefore, under diffusion-limited conditions, pulse charging can enhance charging efficiency, encourage more uniform lithium-ion distribution, and suppress lithium plating.

5.3.4 Constant Temperature-constant Voltage charging

Operating temperature of LIBs evidently influences their performance, safety, and degradation. The conventional CC-CV, advanced MSCC or pulse charging works on explicit control of the battery voltage and current and they do not consider the influence of temperature. The battery temperature is directly influenced by internal resistive losses, electrochemical polarization, and ambient temperatures. Hence, charging the batteries at high C-rates and under extreme temperature conditions may lead to excessive temperature rise and may lead to thermal runaway.

CT-CV charging strategy is an advanced strategy that addresses this limitation by regulating the battery temperatures by actively controlling the charging current [18]. The conceptual depiction CT-CV strategy is as shown in Figure 5.4. The current charge is regulated by using the temperature feedback, the current is reduced when the temperature is higher and the current is increased when the temperature is low, limiting the thermal stress and degradation. Once the maximum temperature is reached during CT phase of charging, the strategy

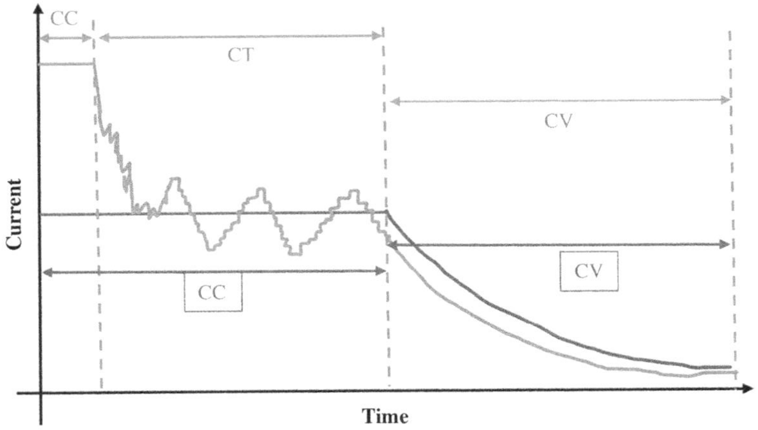

Figure 5.4 Conceptual depiction of CT-CV charging.

changes to CV phase ensuring thermal protection and safe voltage limits during high power charging.

5.4 Charging Standards

EV chargers, charging stations, and other devices that control the transfer of electricity from a power supply to an EV are collectively referred to as electric vehicle supply equipment (EVSE). Although EVSE is the formal industry term, EV chargers, charging stations, charge points, or charging docks are commonly used and widely accepted terms. LIB charging systems are broadly classified as AC charging and DC charging. However, because LIBs are inherently DC devices, the electrical energy must be supplied as direct current. Since the electricity supplied by the grid is an alternating current, an AC-to-DC converter stage is required in all EV charging systems. This conversion can be performed either onboard the vehicle, through an integrated on-board charger (OBC) or offboard within a charging station, known as DC fast chargers capable of supplying higher power to the batteries. The representative architecture of OBC is as shown in Figure 5.5. Different EVSE charging levels are associated with distinct connector types, power outputs, charging speeds, and application scenarios. Depending on the charging level, the AC-to-DC conversion may take place inside the vehicle or within the charging station, thereby influencing charging performance, infrastructure requirements, and system complexity. In addition to providing electrical energy, a charging station is responsible for ensuring safe operation, communication with the vehicle, and adherence to grid and electrical standards. Although EVs are the main purpose of charging stations, stationary BESS uses many of the same architectural principles and operational requirements.

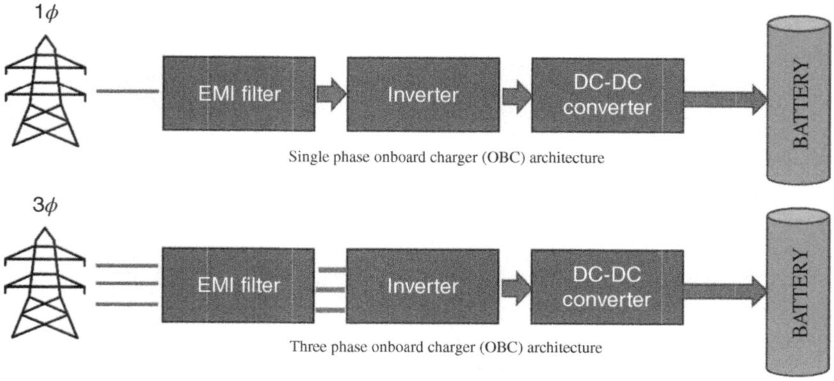

Figure 5.5 Onboard charger architectures.

From a functional standpoint, battery charging approaches can also be broadly classified into conductive and nonconductive methods. Conductive charging relies on a physical electrical connection between the power source and the battery system, as in conventional AC and DC charging. Nonconductive approaches, such as wireless power transfer, enable energy transfer without direct electrical contact by means of electromagnetic coupling. Battery swapping represents a distinct category in which energy transfer is achieved through battery replacement rather than in situ charging. This classification provides a clear conceptual framework for organizing conventional, emerging, and alternative charging solutions discussed in this chapter.

AC charging is usually associated with Levels 1 and 2 charging, which are designed for long-duration charging applications because the charging power is restricted by the rating of the onboard charger. These charging stations are frequently placed in public parking lots, offices, and residential areas where vehicles are left parked for long periods of time. Standardized connectors are necessary for safe conductive power transfer and basic control communication between the car and the charger. While IEC 62196 Types 1 and 2 connections are extensively used in Europe and other nations adopting IEC standards, SAE J1772 defines the most popular AC charging interface in North America. Single- or three-phase AC power transmission at modest current levels appropriate for prolonged charging times is supported by these connectors.

DC fast charging stations bypass the onboard charger by delivering regulated DC power straight to the car battery using external AC-to-DC conversion. This architecture as shown in Figure 5.6 enables substantially higher charging power and is therefore associated exclusively with Level 3 charging. Ten to several hundred kilowatts of electricity can be delivered via DC fast charging stations, enabling quick energy recharge during long-distance travel or frequent usage. DC rapid

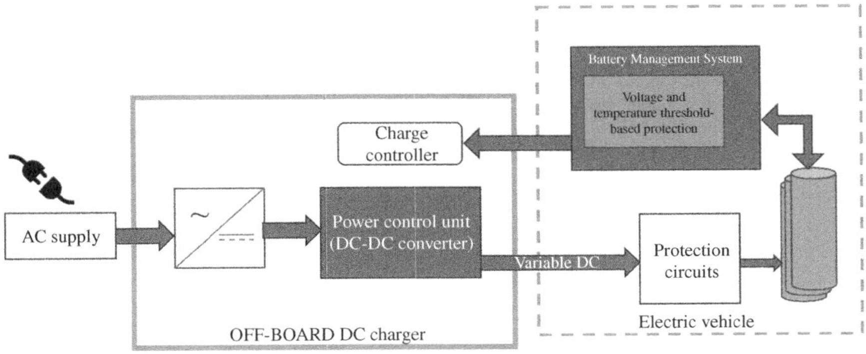

Figure 5.6 Off-board DC-fast charger.

charging necessitates specialized connectors, advanced thermal management, and constant communication with the vehicle BMS to enforce voltage, current, and temperature limits because of the high voltages and currents involved. For DC fast charging, a number of connector standards have been developed, each of which reflects regional preferences and infrastructure development strategies.

In North America and Europe, the combined charging system (CCS) is a popular DC fast-charging interface. The main concept of CCS is that a car can be charged both AC and DC using a single outlet. The CCS connector adds extra high-current DC connectors for DC fast charging in addition to the J1772 AC pins for AC charging. A number of early EV vehicles used CHAdeMO, an earlier DC fast-charging standard that was created in Japan. Although its use has decreased in many areas in comparison to CCS, it is still present in some public infrastructure. Although vehicle-to-grid (V2G) capabilities depends on the entire system design rather than just the connector, CHAdeMO has historically employed CAN-based communication and is frequently addressed in relation to early V2G demonstrations. The GB/T standards in China specify the communication methods and charging connectors that are commonly utilized in that market. GB/T is significant in a global standards chapter because it captures the reality that EV infrastructure has expanded under regional ecosystems, even though it is not the main interface in Canada. By incorporating GB/T, students may better comprehend why the global EV business uses adapters, multi-standard charging stations, and interoperability initiatives. Originally implemented by Tesla, the North American Charging Standard (NACS) is a connector and interface solution that is now available for wider industry use. In keeping with its increasing prominence in North American charging talks, NRCan highlights NACS as a connector that enables both AC and DC charging inside a single physical form factor.

Standard	CHAdeMO	GB/T	CCS type 1	CCS Type 2
Compliance	IEEE 2030.1.1 IEC 62916-3	IEC 62916-3	IEC 62916-3 SAE J1772	IEC 62916-3
Inlet				

Throughout the charging process, constant coordination between the electric car and the charging station is also necessary. While the vehicle BMS imposes cell- and pack-level constraints in real time, the charger must adhere to allowable voltage and current limits throughout operation. Established vehicle–charger communication standards facilitate this synchronization. Among these, ISO 15118 offers high-level digital communication that facilitates sophisticated features like smart charging coordination, automated authentication (plug and charge), and, in some cases, bidirectional charging workflows. Grid-responsive and user-friendly charging operation is made possible by ISO 15118, which permits more extensive interaction between the car and charging station, including the exchange of charging schedules, energy restrictions, and identity credentials, in contrast to basic pilot signaling.

In order to facilitate dependable operation and network-wide coordination, large-scale charging infrastructure necessitates standardized communication between charging stations and backend management systems in addition to direct vehicle–charger communication. Widely used protocols like open charge point protocol (OCPP) provide remote station status monitoring, control command execution (e.g., start and stop charging sessions), configuration management, diagnostics, firmware updates, and transaction handling. Data exchange protocols like Open charge point interface (OCPI), which allow charge point operators and mobility service providers to share information about charger availability, tariffs, user authentication, and session billing, facilitate interoperability across various charging networks at the ecosystem level. These protocols work together to create an operational framework that enables regionally distributed charging infrastructure to operate as a scalable and cohesive system.

Scalable and interoperable charging ecosystems are made possible by charging standards, which go beyond specifying connectors, voltage levels, and communication protocols. Harmonized standards facilitate coordinated operation across manufacturers and regions, reduce system division, and give vehicles access to a variety of charging infrastructure. Intelligent charging behavior, remote monitoring, and integration with energy management systems are further made possible by communication protocols. In this way, charging standards function as fundamental facilitators for future grid integration, operational dependability, and large-scale implementation in addition to being technical specifications.

5.5 Different Charging Levels: Level 1, Level 2, and Level 3 (DC fast charging)

The infrastructure for charging EVs is typically divided into charging levels according to the power, voltage, and current at which energy is supplied to the vehicle. These charge levels offer a useful foundation for comprehending infrastructure needs, appropriate use cases, and charging times. Level 1, Level 2, and Level 3 (DC rapid charging) EV charging are commonly categorized in North America. The fundamental ideas of power level, charging time, and system limitations apply to stationary BESSs, even though this classification is mainly used for EVs. This is especially true when contrasting low- and high-power charging scenarios.

5.5.1 Level 1 Charging

Level 1 charging, which is based on common single-phase home electrical outlets, is the least energy-intensive method of charging EVs. Level 1 charging in North America usually uses 120 V AC and charging currents between 12 and 16 A, producing power levels of about 1–2 kW. Level 1 charging is sometimes regarded as the most accessible charging option for EV customers because it makes use of the electrical infrastructure already present in the home and requires minimal installation work. The Level 1 charging involves supplying low power and hence the battery charging time will be extended [19]. Level 1 charging performs best in situations when cars are parked for long stretches of time, including overnight residential charging or long-term parking situations. Level 1 charging is a comparatively mild charging technique from the standpoint of battery health because the low charging current causes less heat stress and electrochemical deterioration. However, its restricted power supply severely limits its usefulness for fleet-based applications, time-constrained charging, and high-mileage customers.

5.5.2 Level 2 Charging

For use in homes, businesses, and workplaces, Level 2 charging is the most popular type of infrastructure for charging EVs. Power levels range from roughly 3 to 22 kW, depending on the electrical source and charger configuration. It normally operates at 208–240 V AC and can sustain charging currents of up to 80 A. Similar to Level 1 charging, Level 2 charging uses the onboard charger of the car to transform AC power into controlled DC power in order to charge batteries. When compared to Level 1 charging, the higher operational voltage and current drastically shorten the charging time. Level 2 charging is ideal for overnight charging and regular everyday commuting because an average EV battery can be fully charged for 4–10 hours [20]. Level 2 charging provides a workable compromise between battery stress, charging speed, and infrastructure cost. For the majority of LIB chemistries, charging currents typically stay within safe operating limits, even though thermal control may be necessary at the top end of the power range.

As a result, Level 2 charging is extensively used in both public and private charging facilities. Level 2 charging is comparable to moderate-power charging in the context of BESS, when energy is absorbed over several hours as opposed to quickly. This power regime is used by many commercial and grid-scale BESS installations, especially during off-peak hours or when charging from renewable energy sources. This comparison demonstrates how stationary energy storage solutions can benefit from the charging concepts created for EVs.

5.5.3 Level 3 Charging (DC Fast Charging)

Level 3 charging, sometimes referred to as DC fast charging, uses an external AC-to-DC converter within the charging station to deliver high-power direct current straight to the battery pack. By avoiding the vehicle's internal charger, Level 3 charging allows for quick energy replenishment at normal voltages of 400–1000 V and power levels starting at about 50 kW and surpassing 300 kW in ultra-fast charging systems. Under ideal circumstances, 80% SOC can be reached in 20–40 minutes, which is the main advantage of Level 3 charging. However, the high charging power causes significant electrochemical and thermal stress, especially at low temperatures and high SOC, necessitating sophisticated battery management and thermal control.

The choice of charging level involves trade-offs between grid effect, infrastructure cost, battery health, and charging time from the standpoint of system design. While greater charging levels offer comfort at the expense of increasing system complexity and possible degradation, lower charging levels reduce stress and infrastructure complexity but necessitate longer charging times. The battery chemistry and permitted C-rate, thermal management capability, grid-connection limits, and intended use case (private, public, fleet, or grid service) must all be taken into consideration when choosing the charging level in EV and BESS applications. In order to understand charging standards, create charging plans, and carry out useful calculations pertaining to charging time, temperature rise, and charger sizing, it is necessary to comprehend charging levels.

5.6 Emerging EV Charging Approaches

Even though the foundation of today's electric car charging infrastructure is made up of standard conductive charging techniques, such as regular AC charging and DC fast charging, they may not meet all user needs or operating conditions. In order to enhance convenience, decrease charging downtime, or offer flexibility in areas with limited fixed infrastructure, a number of alternative and new charging options have been created to supplement these well-established methods. Battery swapping, wireless power transfer (WPT), and mobile charging are some of these methods that differ from conventional plug-in charging by altering the way energy is supplied to the car or the way charging infrastructure is set up.

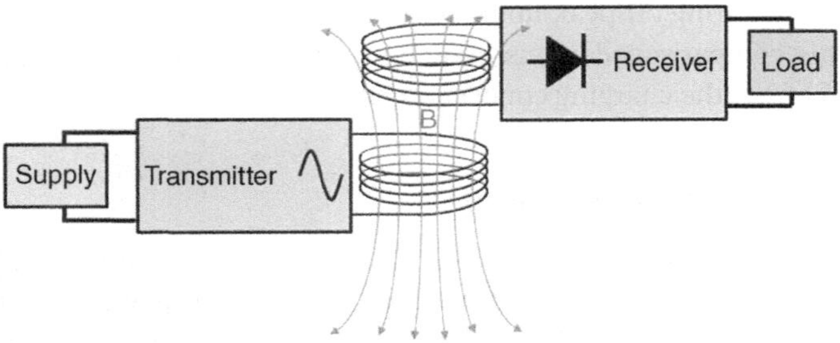

Figure 5.7 Wireless power transfer.

Although these techniques are not yet as popular as traditional charging, they are crucial in certain situations, including fleet operations, confined urban spaces, and emergency or temporary charging situations.

By using electromagnetic coupling to transfer energy over an air gap, WPT for EVs allows battery charging without a physical electrical connection as shown in Figure 5.7. Resonant inductive coupling, in which an alternating current in a ground-side transmitter coil creates a time-varying magnetic field that induces a current in a receiver coil mounted on the vehicle, is the basis for the majority of EV-oriented WPT systems. In order to maintain effective energy transfer under coil misalignment and fluctuating load conditions, power-electronic converters on both sides control frequency, phase, and power flow. In order to facilitate interoperability and the safe deployment of stationary wireless charging systems for passenger cars and light-duty applications, standardization initiatives like SAE J2954 have established operating frequency ranges, power classes, alignment tolerances, and electromagnetic compatibility requirements [21].

Because there are no exposed conductive connections, there is less chance of wear, corrosion, and electric shock, which makes WPT advantageous from a system standpoint in terms of user convenience, environmental robustness, and electrical safety. WPT is especially appealing for applications that require frequent short-duration charging, such shared-mobility platforms, autonomous cars, and taxi fleets, where automated alignment and hands-free operation are advantageous. However, compared to conductive charging, wireless charging systems often show lower efficiency and higher system costs, mainly because of additional power-electronic stages and magnetic coupling losses. Key design issues continue to be thermal control, electromagnetic field exposure limitations, and tolerance to lateral and vertical misalignment.

Battery swapping is an alternative charging approach in which a depleted battery pack is exchanged for a fully charged one at a dedicated swapping station, enabling energy replenishment within a few minutes. Compared with DC fast charging, which typically requires tens of minutes, battery swapping significantly reduces vehicle downtime and allows EVs to be purchased without an onboard battery, lowering initial vehicle cost. Battery ownership, charging, aging, and maintenance are managed centrally by the swapping service provider, reducing user concerns related to battery degradation and lifecycle management [75]. Despite these advantages, battery swapping requires strict standardization of battery form factors, interfaces, and electrical characteristics across vehicle platforms, as well as sufficient inventory management to meet demand. Infrastructure costs associated with battery storage, charging, and thermal control can be substantial, and battery degradation during storage and repeated swapping must be considered [22]. From a system perspective, battery swapping stations can also act as grid-interactive assets, supporting battery-to-grid services and renewable energy integration, making battery swapping an attractive option for both mobility and energy system applications [23].

In contrast to permanent charging infrastructure, mobile charging refers to charging solutions that use portable or vehicle-mounted charging equipment to give electrical energy to an electric car. These systems could be movable power units that can be deployed as needed, trailer-based chargers, or truck-mounted battery packs. In situations like roadside assistance, event-based charging, fleet depots with inadequate infrastructure, or early-stage EV adoption in places without permanent charge stations, mobile charging is mainly meant for short-term, emergency, or location-constrained situations. Compared to fixed charging stations, mobile charging is typically characterized by lesser power capability and greater operating costs, despite the fact that it enables flexibility and quick deployment without requiring significant infrastructure investment. Additional efficiency losses and logistical complexity result from the mobile unit's need to store, transport, and replace energy. Therefore, rather than taking the place of traditional charging infrastructure, mobile charging is better seen as a supplementary option. However, it can be crucial for facilitating fleet operations, increasing resilience in the event of grid failures or infrastructure problems, and boosting charging accessibility.

5.7 Power-electronic Interfaces for Battery Charging

Battery charging systems rely fundamentally on power-electronic converters to regulate voltage, current, and power flow between the electrical grid and the battery. While the previous sections of this chapter focused on charging methods, standards, and charging levels, it is important to briefly outline the role of power electronics in implementing these concepts in practical charging systems. Power-electronic interfaces enable precise control of charging profiles, electrical isolation, compliance with grid requirements, and coordination with battery management and thermal systems. A high-level overview is presented to establish the connection between charging theory and hardware realization.

The primary function of power electronics in battery charging is to convert electrical energy from the grid or another source into a controlled DC form suitable for charging LIBs. This involves regulating charging current and voltage according to predefined strategies such as CC-CV, MCC, pulse charging, or temperature-aware charging. In addition, power-electronic converters provide galvanic isolation, protect the battery from grid disturbances, and ensure safe operation under abnormal conditions such as overvoltage, overcurrent, or fault events. In both EV and BESS applications, charging performance and efficiency are strongly influenced by converter design, including switching devices, control bandwidth, and thermal capability. As charging power increases, converter losses and thermal constraints become increasingly important, directly impacting system size, cost, and reliability.

5.7.1 AC Charging Architecture (Levels 1 and 2)

In Levels 1 and 2 charging, the charger supplies AC power, and the conversion to DC is performed by an onboard charger integrated within the vehicle. A typical AC charging architecture consists of an input EMI filter, an AC–DC rectification stage, and a DC–DC conversion stage that regulates the battery charging current and voltage as shown in Figure 5.8. The onboard charger communicates with the BMS to enforce voltage, current, and temperature limits during charging. The maximum charging power achievable during AC charging is limited by the rating of the onboard charger, which typically ranges from a few kilowatts up to approximately 22 kW. As a result, even if a higher-power AC supply is available, the vehicle's onboard converter ultimately determines the charging rate. Similar AC–DC–DC architectures are also used in some stationary BESS installations, particularly at low to moderate power levels where long charging durations are acceptable.

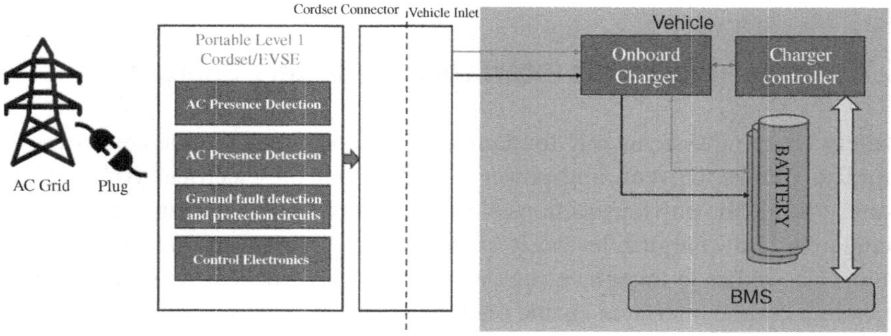

Figure 5.8 Levels 1 and 2 charging.

5.7.2 DC Fast Charging Architecture (Level 3)

Level 3 or DC fast charging uses offboard chargers, where the AC-to-DC conversion is performed entirely within the charging station. In this configuration, regulated DC power is delivered directly to the battery pack, by passing the vehicle onboard charger. A typical DC fast charger consists of a high-power AC-to-DC front-end, followed by an isolated DC-to-DC converter that precisely controls the charging current and voltage. Because DC fast chargers can deliver power levels ranging from tens to hundreds of kilowatts, they enable rapid charging but also introduce significant thermal and electrical stress on the battery. Consequently, DC fast charging requires tight coordination between the charger and the vehicle's BMS, which continuously monitors cell voltages, temperatures, and allowable charging limits. Similar high-power converter architectures are widely used in grid-connected BESS, particularly for applications requiring fast energy absorption or delivery, such as frequency regulation and peak shaving.

5.8 Solved Examples of Battery Charging

This section presents representative numerical examples to illustrate practical calculations related to battery charging. The examples cover charging time estimation, temperature rise during charging, charging rate (C-rate) determination, and charger sizing, which are essential for the design and evaluation of charging systems for EVs and BESS. Unless otherwise stated, idealized assumptions are used to emphasize calculation methodology rather than detailed electrochemical modeling.

5.9 Examples of Charging Time Estimation

Example 1: Charging Time for an EV Using Level 2 Charging

An EV is equipped with a 60-kWh battery pack. The vehicle is charged using a Level 2 charger rated at 7.2 kW, with an overall charging efficiency of 90%. Estimate the time required to fully charge the battery from empty.

Solution: Effective charging power: $P_{\text{eff}} = \eta P_{\text{charger}}$

$$P_{\text{eff}} = 7.2 \text{ kW} \times 0.9$$
$$P_{\text{eff}} = 6.48 \text{ kW}$$

Charging time: $t = \dfrac{E_{\text{battery}}}{P_{\text{eff}}}$

$$t = \frac{60 \text{ kWh}}{6.48 \text{ kW}} = 9.2592 \text{ hours}$$

5.9.1 Charging Time Required is Approximately 9.3 Hours

Example 2: Charging Time for an EV Using CC-CV Charging

An EV is equipped with a 60-kWh battery pack and is charged using a Level 2 charger rated at 7.2 kW. The overall charging efficiency is 90%. Assume that constant-current (CC) charging occurs from 0% to 80% SOC, followed by a constant-voltage (CV) phase from 80% to 100% SOC, during which the average charging power reduces to 50% of the CC power. Estimate the total charging time.

Solution:

1. Effective Charger power during the CC phase

$$P_{\text{CC(eff)}} = \eta \ P_{\text{Charger}}$$
$$P_{\text{CC(eff)}} = 7.2 \text{ kW} \times 0.9$$
$$P_{\text{CC(eff)}} = 6.48 \text{ kW}$$

2. Energy charged during the CC phase (0% to 80% SOC)

$$E_{\text{CC}} = 0.8 \times E_{\text{Battery}}$$
$$E_{\text{CC}} = 0.8 \times 60 \text{ kWh}$$
$$E_{\text{CC}} = 48 \text{ kWh}$$

3. Time required for CC phase

$$t_{\text{CC}} = \frac{E_{\text{CC}}}{P_{\text{CC(eff)}}}$$
$$t_{\text{CC}} = \frac{48 \text{ kWh}}{6.48 \text{ kW}} = 7.41 \text{ hours}$$

4. Effective power during CV phase: Assuming the CV charging power is reduced to 50% of CC power

$$P_{\text{CV}} = 0.5 \ P_{\text{Charger}}$$
$$P_{\text{CV}} = 0.5 \times 6.48 \text{ kW}$$
$$P_{\text{CV}} = 3.24 \text{ kW}$$

5. Energy charged during CV phase (80% to 100% SOC)

$$E_{\text{CV}} = 0.2 \times E_{\text{Battery}}$$
$$E_{\text{CV}} = 0.2 \times 60 \text{ kWh}$$
$$E_{\text{CV}} = 12 \text{ kWh}$$

6. Time required for CV phase

$$t_{\text{CV}} = \frac{E_{\text{CV}}}{P_{\text{CV}}}$$
$$t_{\text{CV}} = \frac{12 \text{ kWh}}{3.24 \text{ kW}} = 3.7 \text{ hours}$$

7. Total charging time:

$$T_{\text{total}} = t_{\text{CC}} + t_{\text{CV}}$$
$$T_{\text{total}} = 7.41 + 3.70 = 11.11 \text{ Hours}$$

Example 3: DC Fast Charging Time (EV)
A 75 kWh EV battery is charged from 20% to 80% SOC using a 150 kW DC fast charger. Assume constant power delivery and 92% efficiency.

Solution: Energy required: $E = (\text{SOC}_{\text{Final}} - \text{SOC}_{\text{Initial}}) \times \text{E}_{\text{rated}}$

$$E = (0.8 - 0.2) \times 75 \text{ kWh}$$
$$E = 45 \text{ kWh}$$

Effective power: $P_{\text{eff}} = \eta P_{\text{charger}}$

$$P_{\text{eff}} = 0.92 \times 150 \text{ kW}$$
$$P_{\text{eff}} = 138 \text{ kW}$$

Charging time:

$$t = \frac{E_{\text{battery}}}{P_{\text{eff}}}$$
$$t = \frac{45}{138} = 0.326 \text{ hours}$$
$$t = 0.326 \times 60 = 19.6 \text{ min}$$

Example 4: Charging Time for a BESS
A grid-connected BESS rated at 500 kWh is charged using a 100 kW charger during off-peak hours. Estimate the charging time assuming 95% efficiency.

Solution:

$$P_{\text{eff}} = \eta P_{\text{charger}}$$
$$P_{\text{eff}} = 0.95 \times 100 = 95 \text{ kW}$$

Charging time:

$$t = \frac{E_{\text{battery}}}{P_{\text{eff}}}$$
$$t = \frac{500 \text{ kWh}}{95 \text{ kW}} = 5.26 \text{ hours}$$

5.9.2 Examples on Temperature Rise During Charging

A simplified lumped thermal model is used:
Heat generated,

$$Q = I^2Rt$$

Temperature rise

$$\Delta T = \frac{Q}{mc_p}$$

Example 5: Temperature Rise During EV Fast Charging
An EV battery is charged at 200 A for 30 minutes. The internal resistance is 20 mΩ, the e battery mass is 400 kg, and the specific heat capacity is 900 J/kg K.

Solution: Heat generated:

$$Q = I^2Rt$$
$$Q = 200^2 \times 0.02 \times 1800$$
$$Q = 1.44 \times 10^6 \text{ J}$$

Temperature rise

$$\Delta T = \frac{Q}{mc_p}$$

$$\Delta T = \frac{1.44 \times 10^6}{400 \times 900} = 4\,°\text{C}$$

Example 6: Temperature Rise in a BESS During Moderate Charging
A BESS module is charged at 80 A for 2 hours with internal resistance 15 mΩ, mass 600 kg, and specific heat 1000 J/kg K.

Solution: Heat generated:

$$Q = I^2Rt$$
$$Q = 80^2 \times 0.015 \times 2 = 7200$$
$$Q = 691.2 \text{ kJ}$$

Temperature rise

$$\Delta T = \frac{691.2\ kJ}{600 \times 1000} = 1.15\ °\text{C}$$

5.9.3 Examples of Charging Rate (C-rate) Determination

Example 7: C-Rate Calculation for an EV Battery
An EV battery has a nominal capacity of 50 Ah and is charged at 100 A. Determine the C-rate.

Solution:

$$C = \frac{I_{\text{charge}}}{Q_{\text{rated}}}$$
$$C = \frac{100}{50} = 2\,C$$

Charging rate is 2C

Example 8: C-Rate for a BESS Cell String
A BESS cell string has a capacity of 200 Ah and is charged at 40 A. Determine the C-rate

Solution:

$$C = \frac{40}{200} = 0.2\,C$$

Example 9: Determining Allowable Charging Current
A lithium-ion cell is rated for a maximum charging rate of 1C and has a nominal capacity of 60 Ah. Determine the maximum allowable charging current.

Solution:

$$I_{\text{max}} = C_{\text{max}} \times Q_{\text{rated}}$$
$$I_{\text{max}} = 1 \times 60 = 60A$$

Maximum allowable charging current is 60A
Answer: Maximum allowable charging current is 60 A.

5.9.4 Examples on Charger Sizing

Example 10: Charger Sizing for Residential EV Charging
An EV is equipped with a 70-kWh battery pack and must be fully charged within 10 hours. Determine the minimum charger power rating required to meet this charging requirement.

Solution:

$$P = \frac{E}{t}$$

$$P = \frac{70}{10} = 7 \text{ kW}$$

Example 11: Fleet Charging Charger Sizing

A depot operates 15 electric buses, each equipped with a 300-kWh battery pack. All buses must be fully charged overnight within an 8-hour window. Determine the total charging power capacity required at the depot, assuming ideal charging conditions.

Solution: Total Energy

$$E_{\text{total}} = 15 \times 300 \text{ kWh} = 4500 \text{ kWh}$$

Required charging power

$$P = \frac{E_{\text{total}}}{t}$$

$$P = \frac{4500}{8} = 562.5 \text{ kW}$$

Example 12: Charger Sizing for a BESS Installation

A grid-scale BESS rated at 1 MWh must be fully charged within 4 hours. Determine the minimum charger power rating required to meet this charging requirement.

Solution: A grid-scale BESS rated at 1 MWh must be charged in 4 hours.

$$P = \frac{1000}{4} = 250 \text{ kW}$$

Including 10% margin, $P_{\text{rated}} = 1.1 \times 250 = 275$ kW

References

1 Y. Li, et al., "Optimized charging of lithium-ion battery for electric vehicles: Adaptive multistage constant current–constant voltage charging strategy," *Renew. Energy*, vol. 146, pp. 2688–2699, 2020.

2 K. Li, F. Wei, K. J. Tseng, and B. H. Soong, "A Practical lithium-ion battery model for state of energy and voltage responses prediction incorporating temperature and ageing effects," *IEEE Trans. Ind. Electron.*, vol. 65, no. 8, pp. 6696–6708, 2018.

3 C. Chen, Z. Wei, and A. C. Knoll, "Charging optimization for Li-ion battery in electric vehicles: A review," *IEEE Trans. Transp. Electrif.*, vol. 8, no. 3, pp. 3068–3089, 2022.

4 S. K. Bhoi and M. R. Nayak, "Optimal scheduling of battery storage with grid tied PV systems for trade-off between consumer energy cost and storage health," *Microprocess. Microsyst.*, vol. 79, 2020, 103274, ISSN 0141-9331, https://doi.org/10.1016/j.micpro.2020.103274. (https://www.sciencedirect.com/science/article/pii/S0141933120304336

5 J. Sarda, N. Patel, H. Patel, R. Vaghela, and B. Brahma, "A review of the electric vehicle charging technology, impact on grid integration, policy consequences, challenges and future trends," vol. 12, pp. 5671–5692, 2024, ISSN 2352-4847, https://doi.org/10.1016/j.egyr.2024.11.047.

6 R. Abousleiman, "Charge capacity versus charge time in CC-CV and pulse charging of Li-ion batteries," 2026.

7 B. M. T. Lawder et al., "System (BESS) and battery management system (BMS) for grid-scale applications," *Proc. IEEE*, vol. 102, no. 6, pp. 1014–1030, 2014.

8 W. Shen, T. T. Vo, and A. Kapoor, "Charging algorithms of lithium-ion batteries: An overview," in *Proceedings of the 2012 7th IEEE Conference on Industrial Electronics and Applications, ICIEA 2012*, 2012, pp. 1567–1572.

9 C. Chetri and S. Williamson, "Health-conscious fast charging for electrified aircraft batteries using a multistage-constant-current temperature-controlled strategy," *IEEE J. Emerg. Sel. Top. Power Electron.*, vol. 13, no. 4, pp. 4635–4642, 2025.

10 G. Ji, L. He, T. Wu, and G. Cui, "The design of fast charging strategy for lithium-ion batteries and intelligent application: A comprehensive review," vol. 377, 2025.

11 MIT Electric Vehicle Team, "*A Guide to Understanding Battery Specifications*," 2008.

12 G. Nuroldayeva, Y. Serik, D. Adair, B. Uzakbaiuly, and Z. Bakenov, "State of health estimation methods for lithium-ion batteries," *Int. J. Energy Res.*, vol. 2023, 2023.

13 W. Mei, L. Zhang, J. Sun, and Q. Wang, "Experimental and numerical methods to investigate the overcharge caused lithium plating for lithium ion battery," *Energy Storage Mater.*, vol. 32, no. April, pp. 91–104, 2020.

14 K. Yu, A. Miracle, S. Choe, and W. Lee, "Optimized multi-stage constant current fast charging protocol suppressing lithium plating for lithium-ion batteries using reduced order electrochemical-thermal-life model," *J. Power Sources*, vol. 626, p. 235759, 2025.

15 F. An, R. Zhang, Z. Wei, and P. Li, "Multi-stage constant-current charging protocol for a high-energy-density pouch cell based on a 622NCM/graphite system," *RSC Adv.*, vol. 9, no. 37, pp. 21498–21506, 2019.

16 F. M. N. Uddin, M. G. Rasul, A. S. M. Sayem, and N. K. Mandal, "Design and optimization of lithium-ion battery as an efficient energy storage device for electric vehicles : A comprehensive review," vol. 71, 2023.
17 Y. Tae et al., "Electrochimica acta insight into pulse-charging for lithium plating-free fast-charging lithium-ion batteries," vol. 462, 2023.
18 L. Patnaik, A. V. J. S. Praneeth, and S. S. Williamson, "A closed-loop constant-temperature constant-voltage charging technique to reduce charge time of lithium-ion batteries," *IEEE Trans. Ind. Electron.*, vol. 66, no. 2, pp. 1059–1067, 2019.
19 J. Sears, D. Roberts, and K. Glitman, "A comparison of electric vehicle level 1 and level 2 charging efficiency," pp. 2012–2015, 2014.
20 S. Khan, S. Shariff, A. Ahmad, and M. Alam, "A comprehensive review on level 2 charging system for electric vehicles a comprehensive review on level 2 charging system for electric vehicles," *Smart Sci.*, vol. 0477, pp. 1–23, 2018, https://doi.org/10.1080/23080477.2018.1488205
21 D. Vincent, S. Member, S. S. Williamson, and S. Member, "Role of dielectrics in the capacitive wireless power transfer system," pp. 1217–1222, 2020.
22 D. Vincent, P. S. Huynh, and S. S. Williamson, "A link-independent hybrid inductive and capacitive wireless power transfer system for autonomous mobility," vol. 3, pp. 211–218, 2022.
23 D. Vincent, P. S. Huynh, and S. S. Williamson, "A novel three leg inverter for high power hybrid inductive and capacitive wireless power transfer system," pp. 1544–1548, 2019.

6

Second-life Applications of Retired Electric Vehicle Batteries

6.1 Introduction

The accelerating global deployment of electric vehicles (EVs) is creating a rapidly growing stream of retired traction batteries. In automotive service, lithium-ion packs are commonly removed from vehicles when their available capacity or power capability drops below mobility requirements, typically around 70–80% of initial capacity, even though the batteries can still deliver meaningful energy and power in less demanding duty cycles [1]. This creates an opportunity for second-life batteries (SLBs), where retired EV packs, modules, or cells are repurposed for stationary or semi-stationary energy storage before final recycling. From an energy-systems perspective, SLBs are compelling because many grid and behind-the-meter services prioritize usable energy, availability, and moderate power response rather than the mass- and volume-constrained power density required for traction. As electrification expands alongside higher penetrations of variable renewable generation, storage is increasingly needed for time shifting, peak shaving, resiliency/backup, renewable smoothing, and other flexibility services. SLBs can contribute to these needs at potentially lower upfront cost, but their value depends on careful matching between remaining capability and application requirements [2, 3].

Second-life deployment also connects to environmental and economic benefits. Life-cycle studies emphasize that EV impacts depend strongly on electricity generation mix and that battery manufacturing is a major contributor to production-phase impacts; extending battery use through second life can improve lifetime utilization and value retention [2, 4]. However, scaling SLBs is challenging as retired batteries exhibit heterogeneity in state of health (SOH), internal resistance, and degradation trajectory, and uncertainty in remaining useful life (RUL) complicates bankable design and warranties. Recent perspectives highlight that fast, reliable diagnostics, robust lifetime prediction under new duty cycles, and standardized processes are decisive for competitiveness in stationary storage markets [1, 3].

Accordingly, this chapter adopts an energy and electrical engineering viewpoint to organize second-life deployment around (i) application taxonomies, (ii) quantitative performance metrics (technical, economic, reliability, and selected environmental indicators), and (iii) structured assessment, grading, and application matching workflows. We emphasize decision-relevant metrics such as usable capacity, power limits linked to resistance, efficiency, thermal headroom, availability, and levelized cost that determine whether SLBs can reliably deliver grid services and behind-the-meter value streams. We also recognize that successful second-life value chains increasingly require coordination among stakeholders across automotive, energy, remanufacturing, and recycling domains, and that emerging circular-ecosystem and circular-supply-chain perspectives are shaping how SLBs are sourced, qualified, and deployed at scale [5, 6].

The remainder of this chapter is organized as follows. Section 6.2 discusses the drivers, challenges, and value proposition of SLBs. Section 6.3 presents an application taxonomy based on mobility level. Section 6.4 defines the key performance metrics and introduces multi-criteria evaluation concepts. Sections 6.5 and 6.6 describe assessment/testing workflows and grading-to-application matching. Section 6.7 summarizes major deployment domains and representative case studies. Section 6.8 discusses emerging trends shaping SLB value chains. Section 6.9 provides additional solved examples.

6.2 Motivation, Challenges, and Value Proposition of Second-life Batteries

Retired EV battery packs typically retain substantial functional capability that can be leveraged in less demanding applications, particularly stationary and grid-connected energy storage. Because traction batteries are commonly retired based on mobility-driven performance thresholds—such as reduced range, diminished peak power capability, or warranty limits—rather than complete loss of functionality, a significant portion of their embedded energy-storage value often remains available for continued service [2, 4, 7].

At the same time, immediate recycling alone is unlikely to absorb the rapidly growing stream of retired batteries in the near to medium term. Practical limitations related to collection logistics, testing capacity, processing throughput, and incomplete recovery efficiencies for certain materials constrain recycling pathways [1, 3]. Within this context, second-life deployment should be viewed not as a substitute for recycling, but as a complementary and transitional strategy that extends functional utilization prior to final material recovery.

6.2.1 Why Second Life is Attractive from an Energy and Electrical Perspective

From an energy-systems standpoint, stationary applications impose fundamentally different constraints than traction. In grid-tied or behind-the-meter environments, energy capacity, availability, and moderate power response are often more critical than mass- and volume-specific power density. Consequently, batteries that no longer meet automotive performance requirements may still satisfy the technical needs of peak shaving, time shifting, renewable smoothing, backup power, and selected ancillary services when operated under conservative derating and appropriate power-electronic interfaces.

The increasing penetration of variable renewable generation and electrified loads further amplifies demand for distributed flexibility resources. SLBs can contribute to this flexibility by providing controllable energy buffering and load shifting. However, the realized value of such deployments is strongly application-dependent and governed by duty-cycle severity, round-trip efficiency, resistance growth, and degradation under the new operating regime [2, 7, 8]. Proper alignment between residual capability and service requirements is therefore essential to ensure both technical feasibility and economic viability.

6.2.2 Key Technical Challenges: Heterogeneity, Uncertainty, and Bankability

Despite their potential, SLBs are inherently heterogeneous assets. Variations in first-life operating history, including cycling intensity, fast-charging exposure, depth-of-discharge patterns, and thermal conditions lead to dispersion in residual capacity, internal resistance, power capability, efficiency, and safety margin. Even modules originating from the same vehicle pack may exhibit non-uniform degradation due to localized thermal gradients or current imbalance.

Aging mechanisms are path-dependent and nonlinear, complicating reliable extrapolation of RUL from limited diagnostic data. Uncertainty in degradation trajectories directly affects electrical and thermal design decisions, including series–parallel regrouping strategies, balancing requirements, current limits, and thermal management provisions. These technical uncertainties translate into financial risk, influencing warranty design, insurance considerations, and project bankability [3, 7–9]. Managing heterogeneity and uncertainty is therefore a central engineering and economic challenge in scaling second-life deployment.

6.2.3 Diagnostics, Grading, and Compliance Considerations

Scalable second-life ecosystems require reliable screening and grading methodologies that connect observed battery condition to enforceable operating limits and credible lifetime projections. Diagnostics must quantify both capacity fade and power-related degradation due to resistance growth, while incorporating uncertainty bounds suitable for engineering and financial decision-making.

Recent studies emphasize that diagnostic accuracy, uncertainty quantification, and robust lifetime modeling are decisive factors for competitiveness in stationary storage markets [1, 3]. In practice, standardized evaluation frameworks are increasingly important. For example, UL 1974 provides procedures for sorting and grading batteries intended for repurposed use, while complementary safety standards address stationary integration requirements [10]. Such standards formalize documentation, traceability, and qualification processes, reducing liability exposure and supporting regulatory acceptance.

6.2.4 Value Proposition Across Environmental, Economic, and Grid Dimensions

The value proposition of SLBs spans environmental, economic, and grid-system dimensions. From a sustainability perspective, extending battery service life increases lifetime utilization of the embedded manufacturing energy and materials, potentially reducing the environmental impact per delivered kilowatt-hour relative to immediate replacement with new storage capacity. The magnitude of this benefit depends on the electricity mix, utilization intensity, and recycling efficiency [2, 4, 11].

Economically, second-life batteries can be competitive for selected services when refurbishment costs, balance-of-system integration, and operating constraints are properly accounted for. However, project viability is sensitive to refurbishment expense, achievable revenue stacking, degradation rate under the intended duty cycle, and the credibility of lifetime projections [7, 9, 12]. Conservative operating windows may enhance durability but reduce usable energy, highlighting inherent trade-offs between short-term revenue and long-term value retention. At the grid level, SLBs can support flexibility needs including buffering, load shifting, congestion mitigation, and localized resilience. When strategically sited and appropriately sized, they may defer infrastructure upgrades or reduce peak demand stress [2, 8].

Overall, SLBs occupy a critical intersection between sustainability objectives, economic opportunity, and grid flexibility requirements. Their deployment is motivated by the opportunity to extract additional electrical value prior to recycling, yet constrained by heterogeneity, degradation uncertainty, diagnostic limitations, and safety and compliance obligations. These considerations motivate the metrics-driven application taxonomy and structured assessment and decision-making frameworks developed in the following sections.

6.3 Taxonomy of Second-life Applications

SLBs can be deployed across a broad set of use cases, but the suitability of a retired EV battery depends strongly on the operating environment, mobility constraints, and duty-cycle stressors including average C-rate demand, depth of discharge (DoD), temperature exposure, and cycling frequency, which jointly govern achievable power limits, efficiency, degradation rate, and uncertainty in RUL [3, 9, 13–15]. To organize this space in a decision-relevant way, this chapter adopts a practical taxonomy based on the degree of mobility of the end-use system.

6.3.1 Classification by Mobility Level

An expired EV battery can be repurposed into three broad mobility categories, low-demand mobile applications, semi-stationary applications, and stationary applications [16, 17]. Figure 6.1 summarizes this classification and provides representative examples. This taxonomy is useful because the degree of mobility is closely linked to a distinct set of design and operational constraints, including mechanical and packaging requirements, exposure to vibration and impact, and variability in operating conditions. Stationary deployments typically permit relaxed space and weight constraints and support sustained energy throughput over long durations, whereas mobile deployments impose stricter limitations on packaging density, thermal robustness, and safety margins due to dynamic mechanical loading and more variable operating environments.

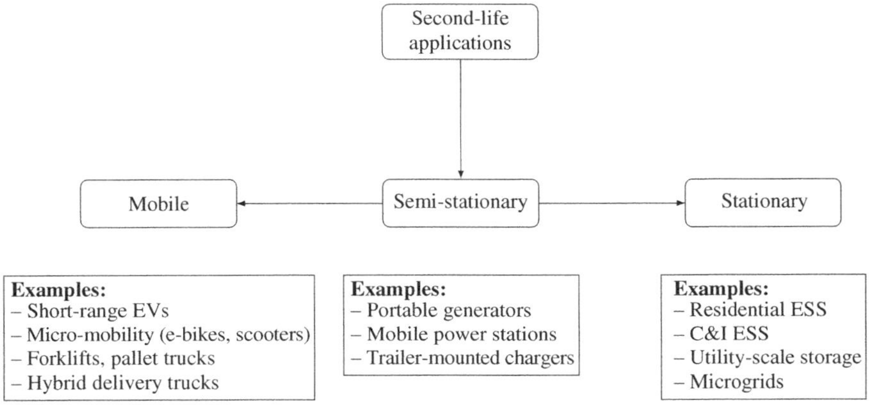

Figure 6.1 Classification of second-life applications grouped by mobility level.

6.3.2 Use Cases and Technical Requirements

The mobility categories described above do not define a single operating pattern for SLBs. Instead, they indicate the types of operating conditions and system requirements that are generally feasible in each application context. Different levels of mobility imply different mechanical stresses, thermal conditions, and space constraints, which in turn affect how much power the battery can safely deliver and how it can be integrated into a system as summarized in Table 6.1. Because SLBs are more diverse in condition and have smaller safety margins than new batteries, applications that require moderate power, operate within controlled temperature ranges, and allow conservative derating are generally better suited for reliable second-life use.

6.3.2.1 Mobile Applications

Mobile SLB applications include short-range or low-utilization electric mobility and industrial vehicles where extreme power density is not the primary constraint. Representative targets include micro-mobility such as e-bikes, e-scooters, light EVs such as golf carts and small utility vehicles, and industrial vehicles such as forklifts and pallet trucks [16, 18, 19]. In some designs, SLBs may also support auxiliary loads such as on-board refrigeration or service electronics, where operating currents can be managed through conservative power limits and supervisory control. The typical requirements for these applications include primarily compact packaging, robust mechanical protection, moderate C-rate capability, conservative thermal management, and strong safety monitoring due to higher exposure to vibration and impact (Figure 6.2) [18, 19].

6.3.2.2 Semi-stationary Applications

Semi-stationary applications refer to systems that remain fixed during operation but can be relocated between sites when needed. Typical examples include portable power generators, mobile power stations supplying temporary or remote loads [16], trailer- or vehicle-mounted energy storage units, and portable EV charging systems. These systems combine the energy density advantages of SLBs with greater deployment flexibility than permanently installed storage.

Portable EV charging represents an important semi-stationary use case and includes both wired charging units and emerging wireless charging platforms.

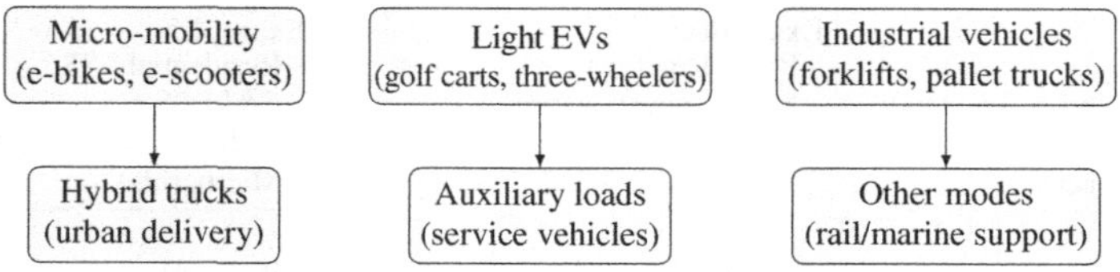

Figure 6.2 Representative mobile second-life applications.

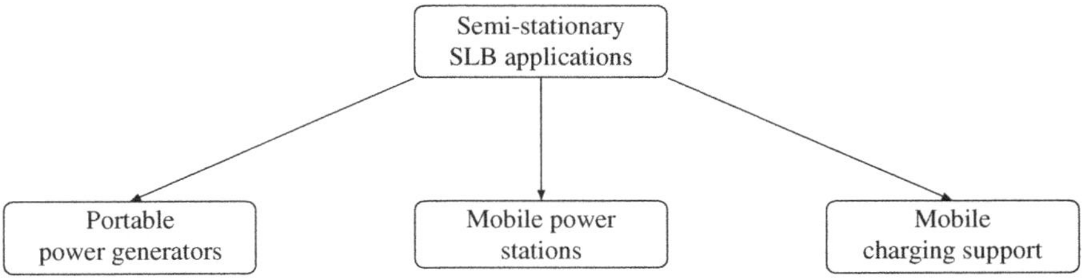

Figure 6.3 Illustrative semi-stationary applications for SLBs.

Wired portable chargers are commonly deployed for temporary charging at construction sites, fleet depots, events, roadside assistance scenarios, or during grid outages [8]. Wireless portable charging systems [20], typically based on inductive power transfer, enable contactless charging for fleet vehicles or controlled parking environments where ease of deployment and reduced mechanical wear are priorities. In both cases, SLBs can provide buffered energy delivery while reducing dependence on grid upgrades or diesel generators. Compared with fully mobile applications, semi-stationary systems experience reduced exposure to continuous vibration and impact, allowing more flexible packaging and enclosure designs. Typical technical requirements include moderate power demand, limited cycling intensity, controlled thermal conditions, and safe, transportable enclosures. Power-electronic interfaces must support site-adaptable operation, grid or off-grid modes, and standardized EV charging protocols, while conservative derating is often applied to account for battery heterogeneity and ensure reliable operation (Figure 6.3).

6.3.2.3 Stationary Applications

Stationary deployments represent the dominant pathway for large-scale SLB utilization because they allow (i) flexible packaging, (ii) more controlled thermal environments, and (iii) straightforward integration with power electronics and supervisory energy management. Accordingly, stationary storage systems are widely identified as the most common and mature second-life use case in the literature and in pilot deployments [1, 3]. Stationary SLB systems are commonly deployed as energy storage systems (ESS) in residential, commercial, and industrial (C&I), and utility settings. Detailed stationary sub-domains, including grid-tied storage, microgrids, renewable, and backup/ uninterruptible power supply (UPS) are treated systematically later in Section 6.7. This type of application typically requires high usable capacity, low-to-moderate C-rate operation, long service life, strong monitoring and protection, maintainability, and compliance with stationary ESS standards.

Table 6.1 Mobility-based taxonomy and typical requirements for SLBs.

Category	Representative use cases	Typical requirements/notes
Mobile	Micro-mobility, light EVs, industrial vehicles, auxiliary loads	Space/weight constraints; higher mechanical exposure; moderate C-rate; conservative thermal/safety design
Semi-stationary	Portable generators, mobile power stations, mobile charging support	Moderate cycling; relocatable enclosures; relaxed mechanical constraints vs. mobile; site-adaptable power interface
Stationary	Residential ESS, C&I ESS, utility-scale storage, microgrids	High capacity; low-to-moderate C-rate; long service life; strong monitoring/protection; compliance with stationary ESS standards

6.3.3 Repurposing Level and Integration Implications

Beyond mobility, second-life feasibility depends on the refurbishment depth at which the battery is repurposed:

- **Pack-level reuse** minimizes disassembly cost and time, but offers limited flexibility in managing cell-to-cell heterogeneity.
- **Module-level reuse** provides a balance between reconfiguration flexibility and cost, and is frequently preferred in practice.
- **Cell-level reuse** offers maximum flexibility for sorting and matching, but typically incurs the highest labor, testing, and validation cost.

In general, minimizing disassembly operations can substantially reduce processing cost, whereas deeper refurbishment can improve performance consistency and safety through tighter matching and selective replacement. These trade-offs motivate the structured assessment, grading, and application-matching methodologies discussed in Sections 6.4–6.5.

6.4 Performance Metrics of Second-life Applications

A rigorous evaluation of SLB applications requires more than a qualitative description of potential use cases; it demands a structured and quantitative set of performance metrics that capture the technical condition, economic viability, environmental implications, and operational reliability of repurposed batteries. Unlike new battery systems, SLBs exhibit heterogeneous SOH, uncertain RUL, and application-dependent performance limits, making standardized metrics essential for objective comparison and informed decision-making. Well-defined metrics

Table 6.2 Overview of key performance metric categories for second-life batteries.

Category	Typical metrics	Purpose
Technical	SOH, capacity retention, DCIR/ACIR, round-trip efficiency, thermal margins	Quantify the ability to deliver energy/power safely over time
Economic	Pack & system cost ($/kWh), salvage value, LCOE/LCOS, payback, NPV, IRR	Assess cost-competitiveness vs. new batteries and other technologies
Environmental	GWP ($kgCO_2eq/kWh$), acidification, eutrophication, resource depletion, circularity indicators	Evaluate life-cycle impact and circular-economy benefits
Reliability	Availability, MTBF, fault incidence, degradation rate, RUL	Characterize operational stability and risk
Application-specific	Demand-charge reduction, energy shifting, autonomy time, service quality	Map SLB capabilities to specific stationary use cases

enable assessment of remaining capacity, power capability, efficiency, safety margins, cost structure, and service reliability, while also supporting evaluation of life-cycle benefits and circularity impacts relative to new storage technologies. Such metrics are critical for quantifying trade-offs among lifetime extension, safety, and cost, and for supporting multicriteria decision frameworks that match retired EV batteries to suitable second-life applications [8, 9]. In this chapter, performance metrics are organized into four main categories: technical performance metrics (Section 6.4.1), economic performance metrics (Section 6.4.2), environmental and sustainability metrics (Section 6.4.3), and reliability and application-specific metrics (Sections 6.4.4 and 6.4.5). Table 6.2 provides an overview of these metric families, which are discussed in detail in the following subsections.

6.4.1 Technical Performance Metrics

6.4.1.1 Capacity and State-of-health Metrics

Accurate estimation of usable capacity and remaining life is essential for second-life deployment, particularly in stationary energy storage systems. SOH is the most widely used indicator for quantifying battery degradation because it provides a compact representation of remaining operational capability [21]. Capacity-based SOH is commonly defined as:

$$\mathrm{SOH}_Q = \frac{Q_{\mathrm{measured}}}{Q_{\mathrm{rated}}} \times 100\%, \tag{6.1}$$

where Q_{rated} is the nominal capacity specified at beginning-of-life and Q_{measured} is the available capacity measured under a defined test protocol. In addition to capacity fade, degradation is often reflected by an increase in internal resistance, which directly impacts power capability, efficiency, and thermal behavior [22]. A resistance-based SOH metric can be defined as:

$$\text{SOH}_R = \frac{R_{\text{BOL}}}{R_{\text{measured}}} \times 100\%, \tag{6.2}$$

where R_{BOL} is the internal resistance at beginning-of-life and R_{measured} is the current measured resistance. As resistance increases with aging, this formulation yields decreasing SOH values and captures power-related degradation effects that may not be visible from capacity measurements alone [8]. Alternatively, power capability can be used directly to define SOH, particularly for applications where peak or sustained power delivery is critical. A power-based SOH metric may be expressed as

$$\text{SOH}_P = \frac{P_{\text{available}}}{P_{\text{rated}}} \times 100\%, \tag{6.3}$$

where P_{rated} is the rated power at beginning-of-life and $P_{\text{available}}$ is the maximum allowable power under voltage, current, and thermal constraints at the current aging state. Power-based SOH metrics are especially relevant for second-life applications because increased internal resistance and thermal limitations often constrain usable power before significant capacity loss occurs. In practice, comprehensive SOH assessment for SLBs combines capacity-, resistance-, and power-based indicators to account for heterogeneous aging behavior and application-specific performance requirements [23].

Accurate measurement and prediction of SOH are critical for the safe and reliable deployment of SLBs, as errors in SOH estimation can lead to misinterpretation of available performance, incorrect fault diagnosis, and elevated safety risk. In practice, SOH cannot be directly measured and is therefore inferred from observable quantities such as voltage, current, temperature, and impedance, most commonly through indicators of capacity fade and internal resistance growth. This inference is inherently challenging because battery degradation is strongly nonlinear, history-dependent, and influenced by time-varying operating and environmental conditions, while many existing SOH estimation methods rely on offline datasets or fixed operating assumptions that limit their robustness when applied to SLBs. As a result, recent research has emphasized the use of physically meaningful and statistically robust health indicators (HIs) that correlate reliably with SOH under heterogeneous aging conditions. Commonly used HIs include resistance-related metrics derived from DCIR, ACIR, or electrochemical impedance spectroscopy, capacity-related features extracted through incremental capacity analysis, and voltage–current features obtained from standardized cycling or operational data

[24, 25]. For specific chemistries such as lithium iron phosphate (LFP), features such as the capacitance peak and the voltage at peak capacitance have demonstrated strong correlation with degradation state and have been shown to enable highly accurate capacity and RUL estimation, making them particularly attractive for SLB screening and prognostics.

6.4.1.2 Power Performance and Internal Resistance Metrics

The power capability of a battery is fundamentally governed by its internal resistance, which determines voltage drop, efficiency, heat generation, and allowable current under operational constraints. As batteries age, internal resistance typically increases, leading to reduced peak power, increased losses, and heightened thermal stress. For SLBs, internal resistance metrics are therefore central to evaluating suitability for power-dependent applications such as grid support, backup power, and fast-response energy storage [23, 26]. Two complementary resistance metrics are commonly used in practice: direct current internal resistance (DCIR) and alternating current internal resistance (ACIR).

DCIR characterizes the instantaneous ohmic and polarization response of the battery under a current step and is typically evaluated at multiple states of charge (SoC) to capture operating-condition dependence. In a representative DCIR measurement protocol, the cell is initialized at a defined SoC and subjected to a short current pulse of known magnitude I. The resulting voltage change ΔV is measured after a short duration, and the DC internal resistance is computed as:

$$R_{\mathrm{DC}}(\mathrm{SoC}) = \frac{\Delta V}{I}. \tag{6.4}$$

By repeating this procedure across the usable SoC range and over the battery's lifetime, a resistance profile $R_{\mathrm{DC}}(\mathrm{SoC})$ is obtained. The evolution of this profile with aging provides a direct indicator of power capability loss and is widely used in battery management systems and second-life screening workflows [8]. ACIR, in contrast, is derived from electrochemical impedance spectroscopy (EIS) and captures the frequency-dependent dynamic response of the battery. In EIS measurements, a small-amplitude sinusoidal excitation is applied over a range of frequencies, and the resulting complex impedance

$$Z(f) = Z'(f) + jZ''(f) \tag{6.5}$$

is recorded. The AC internal resistance is commonly defined as the magnitude of the impedance at a selected reference frequency f_0, typically in the high-frequency region where inductive effects are negligible:

$$R_{\mathrm{AC}} = |Z(f_0)|, \quad f_0 = 1\ \mathrm{kHz}. \tag{6.6}$$

ACIR metrics are often visualized using Nyquist plots and are used to parameterize equivalent-circuit models, assess charge-transfer and diffusion effects, and support diagnostics of aging and fault conditions [26]. In second-life evaluation, DCIR and ACIR serve complementary roles: DCIR directly reflects usable power limits under realistic load transients, while ACIR provides deeper insight into internal degradation and model-based performance prediction. Together, these metrics form a robust foundation for assessing power performance, derating strategies, and application compatibility of repurposed EV batteries. Figure 6.4 illustrates how SOH degradation increases internal resistance, thereby constraining power capability and efficiency while tightening thermal and safety limits.

6.4.1.3 Efficiency and Cycling Metrics

To quantify degradation progression across both first-life and second-life phases, batteries are subjected to standardized cycling-and-characterization protocols. A typical second-life test may involve:

- cycling in batches of 100 cycles at 30% depth-of-discharge (DoD),
- followed by characterization, including:
  - CC–CV capacity measurement through charge at 0.5C–4.2 V, discharge at 0.2C–2.65 V,
  - DCIR profiling at several SoC points,
  - ACIR/EIS measurement across a chosen frequency range.

From such tests, key metrics include:

- *Round-trip efficiency* (RTE):

$$\eta_{\mathrm{RTE}} = \frac{E_{\mathrm{discharged}}}{E_{\mathrm{charged}}} \times 100\ [\%], \tag{6.7}$$

- *Capacity retention* after N cycles:

$$\mathrm{CR}(N) = \frac{Q(N)}{Q(0)} \times 100\ [\%], \tag{6.8}$$

- *Per-cycle capacity fade* $\Delta Q/Q(0)$ and *per-year capacity fade*,
- *Evolution of DCIR/ACIR* as a function of cycle number.

Example 6.1: *(capacity fade metric)*

Suppose a cell has a rated capacity of 3.4 Ah. After 500 cycles, the measured capacity is 2.9 Ah. Then the SOH and average per-cycle capacity fade are:

$$\mathrm{SOH}_Q(500) = \frac{2.9}{3.4} \approx 85.3\%, \quad \Delta Q_{\mathrm{avg}} \approx \frac{3.4 - 2.9}{500} = 0.001\ \mathrm{Ah/cycle}.$$

6.4.1.4 Thermal and Safety Performance

Thermal and safety performance are critical determinants of SLB viability, particularly in stationary energy storage systems where cells are densely packed, cooling is constrained, and fault propagation can have severe consequences. As lithium-ion batteries (LIBs) age, increases in internal resistance, loss of thermal uniformity, and degradation of protective materials can elevate operating temperatures and reduce safety margins. Consequently, thermal behavior and safety resilience must be rigorously evaluated before repurposed batteries can be deployed in grid support, behind-the-meter storage, or backup power applications [19, 23].

Thermal performance is commonly quantified through a set of operating and abuse-related metrics that capture both steady-state and transient behavior. Key indicators include the allowable operating temperature range $(T_{\mathrm{min}}, T_{\mathrm{max}})$, the average cell or module temperature during cycling $\overline{T}$, and spatial temperature non-uniformity within a module or pack. The latter is often expressed as a maximum temperature gradient

$$\Delta T_{\mathrm{cell}} = \max_i(T_i) - \min_i(T_i), \tag{6.9}$$

where T_i denotes the temperature of the ith cell or sensor location. Elevated ΔT_{cell} values are undesirable, as they accelerate differential aging and can promote localized thermal stress.

Another critical metric is the maximum temperature rise under worst-case operating conditions, typically associated with high C-rate charge/discharge, fault-induced currents, or degraded cooling performance. For stationary systems, this rise is closely linked to internal heat generation and the effective thermal resistance of the enclosure. In parallel, safety evaluation focuses on the thermal runaway (TR) onset margin, defined as:

$$M_{\mathrm{TR}} = T_{\mathrm{onset}} - T_{\mathrm{op,max}}, \tag{6.10}$$

where T_{onset} is the temperature at which rapid self-heating or runaway is initiated under abuse conditions, and $T_{\mathrm{op,max}}$ is the maximum credible operating temperature in the target application. Experimental studies report TR-related transitions typically occurring in the range of 150–200 °C, depending on chemistry, cell format, and test protocol [27, 28]. Aging effects in SLBs can reduce this margin by lowering T_{onset} and increasing $T_{\mathrm{op,max}}$ through higher resistive heating.

Thermal and safety characterization of SLBs is typically conducted through a combination of instrumented cycling in temperature-controlled chambers, spatial temperature monitoring using surface or embedded sensors, and standardized abuse tests such as external short-circuit, overcharge, and external heating. These procedures align with international safety standards for industrial and stationary LIBs, including IEC 62619 and UL 1973, which emphasize containment, fault detection, and mitigation effectiveness [29, 30].

For SLBs, which have already experienced partial aging and usage heterogeneity, maintaining sufficient thermal uniformity and safety margins is a prerequisite for deployment. Without adequate thermal headroom and robust protection behavior, any projected economic or environmental benefit of SLBs is undermined by accelerated degradation and elevated safety risk.

Example 6.2: *(Maximum temperature rise under worst-case load)*

A SLB module operating in a stationary storage system is expected to generate a worst-case heat loss of $P_{\text{loss}} = 160$ W during peak discharge. The measured module-to-ambient thermal resistance, accounting for enclosure and airflow, is $R_\theta = 0.25$ °C/W. Assuming an ambient temperature of $T_a = 25$ °C, the steady-state temperature rise can be approximated as

$$\Delta T_{\text{max}} = P_{\text{loss}} R_\theta = 160 \times 0.25 = 40\ °\text{C}. \tag{6.11}$$

The corresponding module temperature is therefore

$$T_{\text{mod}} = T_a + \Delta T_{\text{max}} = 65\ °\text{C}. \tag{6.12}$$

If the system design limit is 60 °C, this operating point exceeds the allowable threshold, indicating the need for current derating or improved thermal management.

Example 6.3: *(Thermal runaway onset margin in aged cells.)*

Consider a lithium-ion cell chemistry with a reported thermal runaway onset temperature of $T_{\text{onset}} = 150$ °C under accelerated rate calorimetry conditions [28]. A stationary SLB application limits the maximum operating temperature to $T_{\text{op,max}} = 55$°C. The nominal thermal runaway margin is therefore

$$M_{\text{TR}} = 150 - 55 = 95\ °\text{C}. \tag{6.13}$$

If aging effects reduce the effective onset temperature by 15 °C, a conservative assumption supported by large-format cell studies [27], the revised margin becomes

$$M_{\text{TR,aged}} = 135 - 55 = 80\ °\text{C}. \tag{6.14}$$

This reduction highlights the importance of accounting for aging-induced safety degradation when defining thermal limits and screening criteria for second-life deployment.

6.4.1.5 Usable Capacity

Usable capacity is the amount of energy that can be safely extracted under allowable operating constraints, including SoC limits, temperature limits, and power limits. It is typically lower than nominal capacity, especially for SLBs that are derated to enhance safety and extend lifetime. If

- Q_{rated} is the rated capacity (Ah),
- SOH_Q is the capacity-based state of health (fraction),
- $\text{DoD}_{\text{allowed}}$ is the allowed depth-of-discharge (fraction),

then:

$$Q_{\text{usable}} = Q_{\text{rated}} \times \text{SOH}_Q \times \text{DoD}_{\text{allowed}}. \tag{6.15}$$

If V_{nom} is the nominal cell voltage, the corresponding usable energy is:

$$E_{\text{usable}} = Q_{\text{usable}} \times V_{\text{nom}}. \tag{6.16}$$

Example 6.4: Consider $Q_{\text{rated}} = 100$ Ah, $\text{SOH}_Q = 0.75$, $\text{DoD}_{\text{allowed}} = 0.80$, and $V_{\text{nom}} = 3.7$ V. Then:

$$Q_{\text{usable}} = 100 \times 0.75 \times 0.80 = 60 \text{ Ah}, \; E_{\text{usable}} = 60 \text{ Ah} \times 3.7 \text{ V} = 222 \text{ Wh}.$$

Even though the original rated energy was 100 × 3.7 = 370 Wh, only 222 Wh is used in second life due to ageing and conservative operating limits.

6.4.1.6 Calendar Life Metrics

Calendar life refers to degradation that occurs with time even without cycling, driven mainly by storage temperature, SoC, and chemistry-dependent ageing kinetics. For SLBs, which may be stored for long periods or experience low utilization, calendar ageing can be significant. A simple empirical model for calendar capacity loss is:

$$\Delta Q_{\text{calendar}}(t) = k\,t^n, \tag{6.17}$$

where k is a rate constant which depends on temperature and SoC, operating time (t), and n is an empirical exponent, often in the range 0.3–0.7 for LIBs. Temperature dependence is frequently captured by an Arrhenius-type relation:

$$k(T) = k_0 \exp\left(-\frac{E_a}{RT}\right), \tag{6.18}$$

where k_0 is a pre-exponential factor, E_a is the activation energy, R is the gas constant, and T is the absolute temperature.

Example 6.5: Assume $k = 0.03$ (per$\sqrt{\text{year}}$), $n = 0.5$, storage time $t = 4$ years, and initial SOH at start of storage is 90%. Then:

$$\Delta Q_{\text{calendar}} = 0.03 \times (4)^{0.5} = 0.06 = 6\%, \text{SOH}_{\text{new}} = 90\% - 6\% = 84\%.$$

Thus, even without cycling, a substantial capacity loss can occur over multiyear storage periods.

6.4.1.7 Cycle Life Metrics

Cycle life quantifies how many charge–discharge cycles a battery can complete before reaching a defined end-of-life threshold (often 70–80% SOH). For SLBs, reduced depth-of-discharge (DoD) is commonly used to extend remaining cycle life. A simple empirical relation is:

$$N_{\text{cycles}} \propto \frac{1}{(\text{DoD})^k}, \tag{6.19}$$

where k is an empirical exponent. For two DoD conditions:

$$\frac{N_2}{N_1} = \left(\frac{\text{DoD}_1}{\text{DoD}_2}\right)^k, \tag{6.20}$$

where N_1 is the cycle life at DoD_1 and N_2 is the cycle life at DoD_2.

Example 6.6: Suppose the cycle life at 100% DoD is $N_1 = 1000$ cycles, $k = 1$, and second-life operation uses 50% DoD. Then:

$$N_2 = 1000\left(\frac{100}{50}\right)^1 = 2000 \text{ cycles.}$$

Reducing DoD from 100% to 50% approximately doubles the expected cycle life, motivating conservative SLB derating.

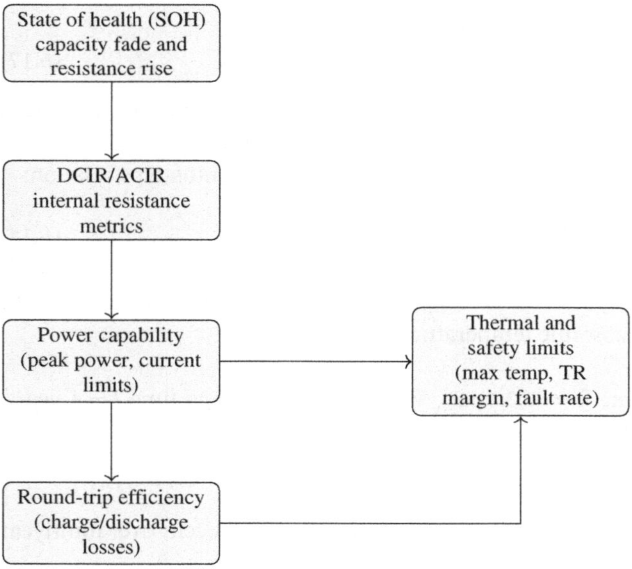

Figure 6.4 Conceptual relationship between key technical metrics in SLBs: SOH degradation influences internal resistance (DCIR/ACIR), which constrains power capability and impacts efficiency, ultimately tightening thermal and safety limits.

6.4.2 Economic Performance Metrics

The economic viability of SLBs depends on their ability to deliver cost-competitive energy and services relative to new battery systems and alternative flexibility options. Economic performance metrics, therefore, aim to quantify cost formation, lifetime value creation, and profitability across applications, while explicitly accounting for ageing, refurbishment effort, and uncertainty in future revenues [4, 8].

6.4.2.1 Cost Structure and Market Price Formation

The market price of SLB systems is governed by a multistage cost structure that extends beyond the acquisition of retired EV battery packs. At minimum, this structure includes the purchase price at EV end-of-life, logistics and transportation, testing and diagnostics, refurbishment and repackaging, balance-of-system components, and installation. These cumulative costs determine the effective pack-level and system-level prices of SLBs and explain the wide spread observed in reported values across studies and applications [23].

Two fundamental cost metrics are used to characterize this structure:

$$C_{\text{pack}}\left[\$/\text{kWh}\right],\ C_{\text{system}}\left[\$/\text{kWh}\right], \tag{6.21}$$

where C_{pack} refers to the refurbished battery pack, including basic casing, cabling, and embedded BMS/thermal management, and C_{system} includes all balance-of-system components and installation.

6.4.2.1.1 Numerical Example

If an SLB pack is refurbished at a cost of 150 \$/kWh and the additional inverter, protection, housing, and installation costs amount to 90 \$/kWh, the resulting system-level cost is

$$C_{\text{system}} = 150 + 90 = 240\ \$/\text{kWh},$$

which can be directly compared against new BESS prices in the range of 400–450 \$/kWh reported in recent studies [31].

6.4.2.2 Levelized Cost of Storage

The levelized cost of storage (LCOS) is a central metric for comparing SLB systems with alternative storage technologies on a lifetime basis. LCOS converts all discounted costs incurred over the project lifetime into an effective cost per unit of electrical energy delivered by the storage system. Unlike upfront cost metrics, LCOS explicitly accounts for degradation, refurbishment, operating expenses, and utilization intensity.

LCOS is defined as:

$$\mathrm{LCOS} = \frac{\sum_{t=0}^{T} \frac{C_{\mathrm{cap},t} + C_{\mathrm{O\&M},t} + C_{\mathrm{rep},t} + C_{\mathrm{end},t}}{(1+r)^t}}{\sum_{t=0}^{T} \frac{E_{\mathrm{delivered},t}}{(1+r)^t}}, \tag{6.22}$$

where r is the discount rate, T is the project lifetime, and $E_{\mathrm{delivered},t}$ is the energy delivered in year t.

Example 6.7: Consider an SLB system with a discounted lifetime cost of 1.5 M$ and a discounted lifetime delivered energy of 10 GWh. The resulting LCOS is

$$\mathrm{LCOS} = \frac{1.5 \times 10^6}{10 \times 10^6} = 0.15 \ \$/\mathrm{kWh}.$$

This value can be benchmarked against wholesale electricity prices, retail tariffs, or LCOS values of new battery systems.

6.4.2.3 Salvage Value and Health-factor-based Pricing

The salvage value of an EV battery at retirement represents the maximum economically justified price at which the battery can enter a second-life market. Salvage value reflects remaining usable life, refurbishment cost, recycling credit, and market acceptance of used products [8]. A commonly used formulation expresses salvage value as:

$$S = \max\left(K_u K_h C_n - C_{rep},\ C_{rc}\right), \tag{6.23}$$

where C_n is the cost of an equivalent new battery, K_h is the health factor, K_u is the used-product discount factor, C_{rp} is the refurbishment cost, and C_{rc} is the recycling credit.

The health factor K_h is often defined as the ratio of the net present value of the lifetime energy throughput of the used battery to that of a new battery operating under identical conditions:

$$K_h = \frac{\mathrm{PVT}_U}{\mathrm{PVT}_N}. \tag{6.24}$$

Example 6.8: Assume $C_n = 400$ \$/kWh, $K_h = 0.6$, $K_u = 0.8$, $C_{rp} = 40$ \$/kWh, and $C_{rc} = 20$ \$/kWh. The salvage value is

$$S = \max(0.8 \times 0.6 \times 400 - 40, 20) = 152 \ \$/\mathrm{kWh}.$$

This value defines an economically rational upper bound for SLB acquisition price.

6.4.2.4 Influence of Depth-of-discharge, Ageing, and Refurbishment

Economic performance in the second life is strongly influenced by operating DoD, cycling frequency, and ageing rate. Higher DoD accelerates degradation, reduces RUL, and lowers the health factor K_h, thereby decreasing salvage value and increasing LCOS [23]. Refurbishment cost further compresses the economically viable price window for SLBs.

An effective metric linking degradation to economics is the cost per full-equivalent cycle:

$$C_{\text{cycle}} = \frac{C_{\text{system}}}{N_{\text{FEC}}}, \tag{6.25}$$

where N_{FEC} is the number of full-equivalent cycles achievable in second life.

Example 6.9: If an SLB system costs 240 \$/kWh and is expected to deliver 3000 full-equivalent cycles, the effective cost per cycle is

$$C_{\text{cycle}} = \frac{240}{3000} = 0.08 \text{ \$/kWh/cycle.}$$

6.4.2.5 Profitability and Payback Time

Profitability metrics quantify how quickly an SLB system can recover its initial investment through operational revenues. A widely used indicator is the simple payback time, defined as the ratio of total upfront investment to average annual net cash flow [4]:

$$\text{Payback Time} = \frac{C_{\text{SLB}} + C_{\text{inst}}}{\overline{C}_{\text{annual}}}. \tag{6.26}$$

Example 6.10: For an SLB system with a total installed cost of 480,000 \$ and an average annual net cash flow of 80,000 \$, the payback time is

$$\text{Payback Time} = \frac{480{,}000}{80{,}000} = 6 \text{ years.}$$

6.4.2.6 Impact on EV Battery Upfront Cost

Second-life utilization can reduce the effective upfront cost of EV batteries by allowing manufacturers or owners to recover a portion of the battery value at end of life. This effect can be expressed as a discounted salvage credit [8]:

$$D = Se^{-\alpha t}, \tag{6.27}$$

where S is the salvage value, α is the discount rate, and t is the ownership period.

Example 6.11: If $S = 150$ \$/kWh, $\alpha = 0.08$, and $t = 8$ years, the effective credit at purchase is

$$D = 150e^{-0.08\times 8} = 67 \text{ \$/kWh}.$$

This corresponds to a reduction of approximately 17% for a new battery priced at 400 \$/kWh.

6.4.2.7 Uncertainty and Market Absorption Potential

Economic projections for SLBs are subject to uncertainty arising from battery price trajectories, electricity market evolution, regulatory frameworks, and the scale and timing of EV battery retirement. Consequently, economic performance is best reported using ranges of LCOS, salvage value, and payback time under multiple scenarios [31]. These metrics enable assessment of market absorption potential and alignment between future SLB supply and realistic demand across stationary applications.

6.4.3 Environmental and Sustainability Performance Metrics

Beyond economic considerations, a primary motivation for SLB deployment is the potential environmental benefit of extending battery use before recycling or disposal. By reusing EV batteries in stationary applications, second life supports circular economy principles by spreading the environmental burden of battery manufacturing over a larger amount of delivered energy, delaying material-intensive recycling processes, and displacing alternative high-emission flexibility options such as fossil-fueled peaking plants [4, 31].

6.4.3.1 Greenhouse Gas Emissions and Life-cycle Impact Metrics

Environmental performance of SLBs is most commonly evaluated using life cycle assessment (LCA), which quantifies environmental impacts over the entire life cycle of a product or system. The most widely reported metric is global warming potential (GWP), expressed in kilograms of CO_2 equivalent per unit of useful energy delivered. For battery systems, GWP is typically normalized either per delivered kilowatt-hour or per kilometer of vehicle operation [32, 33].

The GWP per unit of delivered energy can be expressed as:

$$\text{GWP}_{\text{del}} = \frac{E_{\text{prod}} + E_{\text{rep}} + E_{\text{EoL}} + E_{\text{op}}}{E_{\text{delivered}}}, \tag{6.28}$$

where E_{prod} denotes emissions from battery production, E_{rep} emissions from refurbishment or replacement, E_{EoL} emissions associated with end-of-life treatment, and E_{op} operational emissions associated with electricity losses and auxiliary consumption. LCA studies consistently report that second-life use of EV batteries in stationary applications such as peak shaving or renewable capacity firming can reduce GWP by approximately 20–30% compared to deploying new batteries for the same service, even when additional handling and refurbishment emissions are included [4, 31].

Example 6.12: Assume a battery system with a combined production, refurbishment, and end-of-life emission of 6000 kg CO_2-eq delivers 40 MWh of energy over its first and second life. The resulting GWP per delivered kilowatt-hour is

$$\mathrm{GWP}_{\mathrm{del}} = \frac{6{,}000}{40{,}000} = 0.15 \text{ kg } \mathrm{CO_2\text{-}eq/kWh}.$$

If the same stationary service were provided by a new battery with higher production emissions, the GWP per delivered kWh would increase accordingly.

6.4.3.2 Combined First-life and Second-life Environmental Benefits

Environmental assessments of SLBs often evaluate the combined benefit of EV operation followed by stationary reuse. This approach accounts for emissions avoided during first-life vehicle operation relative to internal combustion engine (ICE) vehicles, as well as emissions avoided in second life by displacing alternative grid assets [8].

The combined life-cycle GWP reduction can be expressed as

$$\Delta\mathrm{GWP}_{\mathrm{total}} = \Delta\mathrm{GWP}_{\mathrm{EV}} + \Delta\mathrm{GWP}_{\mathrm{SLB}}, \tag{6.29}$$

where $\Delta\mathrm{GWP}_{\mathrm{EV}}$ represents avoided emissions from vehicle electrification and $\Delta\mathrm{GWP}_{\mathrm{SLB}}$ represents avoided emissions from second-life displacement of new storage or fossil-based peaking generation. Multiple studies show that pairing EV adoption with second-life stationary applications can approximately double the total GWP reduction compared to vehicle electrification alone, particularly when SLBs replace gas peaker plants or new battery installations [31, 34].

Example 6.13: If EV electrification avoids 25 t CO_2-eq over the vehicle lifetime and second-life deployment avoids an additional 20 t CO_2-eq by replacing a gas peaker, the combined benefit is

$$\Delta\mathrm{GWP}_{\mathrm{total}} = 25 + 20 = 45 \text{ t } \mathrm{CO_2\text{-}eq},$$

corresponding to an 80% increase relative to vehicle electrification alone.

6.4.3.3 Degradation, Efficiency, and Environmental Trade-offs

The environmental benefit of SLBs is influenced by degradation behavior and system efficiency. Capacity fade and reduced round-trip efficiency increase the amount of electricity required to deliver a given amount of useful energy, thereby increasing operational emissions per delivered kilowatt-hour [35].

This effect can be captured by defining an effective emission intensity that accounts for round-trip efficiency η_{rt}:

$$\mathrm{GWP}_{\mathrm{eff}} = \frac{\mathrm{GWP}_{\mathrm{base}}}{\eta_{\mathrm{rt}}}, \tag{6.30}$$

where $\mathrm{GWP}_{\mathrm{base}}$ is the baseline emission intensity of the supplied electricity.

Example 6.14: If grid electricity has an emission intensity of 0.40 kg CO_2-eq/kWh and the SLB system operates at a round-trip efficiency of 80%, the effective emission intensity becomes

$$\mathrm{GWP}_{\mathrm{eff}} = \frac{0.40}{0.80} = 0.50 \text{ kg } CO_2\text{-eq/kWh}.$$

This illustrates how efficiency losses can partially offset environmental gains, particularly in regions with carbon-intensive electricity.

6.4.3.4 Second Life Versus Recycling and Circular Economy Pathways

Second-life reuse and recycling represent complementary strategies within a circular battery economy. Second life extends the functional lifetime of batteries and delays energy-intensive recycling processes, while recycling enables recovery of critical materials such as lithium, nickel, cobalt, copper, and aluminum [36]. The environmental trade-off between these pathways depends on recycling efficiency, energy intensity, and the marginal environmental benefit of recovered materials.

From a life-cycle perspective, the environmental advantage of the second life can be expressed as avoided production impact per delivered energy:

$$\Delta\mathrm{GWP}_{\mathrm{avoid}} = \frac{E_{\mathrm{prod,new}} - E_{\mathrm{prod,SLB}}}{E_{\mathrm{delivered,SLB}}}, \tag{6.31}$$

where $E_{\mathrm{prod,new}}$ is the production emission of a new battery system providing the same service.

Example 6.15: If manufacturing a new battery emits 120 kg CO_2-eq/kWh and refurbishing a SLB emits 30 kg CO_2-eq/kWh, the avoided production emission is 90 kg CO_2-eq/kWh. When spread over 600 kWh of delivered second-life energy, this corresponds to an avoided impact of

$$\Delta\mathrm{GWP}_{\mathrm{avoid}} = \frac{90}{600} = 0.15 \text{ kg } \mathrm{CO}_2\text{-eq/kWh}.$$

Overall, current evidence indicates that for many near- and medium-term stationary applications, second-life deployment provides a robust environmental benefit relative to both immediate recycling and new battery production. As recycling technologies mature and scale, the relative advantage of second life may evolve, reinforcing the need for dynamic, scenario-based sustainability metrics.

6.4.4 Reliability and Operational Performance Metrics

The long-term viability of SLB systems depends not only on cost and environmental benefit, but also on their ability to deliver predictable and dependable performance under real-world operating conditions. Reliability and operational performance metrics quantify how consistently repurposed batteries can provide their intended services despite accumulated first-life ageing, heterogeneous cell histories, and increased uncertainty in degradation behavior. These metrics are particularly critical in grid-support, backup power, and microgrid applications, where unplanned downtime or performance loss can lead to service interruptions or financial penalties [8, 31].

6.4.4.1 Availability and Uptime

System availability measures the proportion of time during which an SLB system is capable of delivering its contracted service at the required power and energy levels. High availability is essential for applications such as frequency regulation, reserve provision, and backup power, where service commitments are time-critical [37].

Availability A over a given observation period can be defined as:

$$A = \frac{T_{\text{operational}}}{T_{\text{total}}}, \tag{6.32}$$

where $T_{\text{operational}}$ is the cumulative time during which the system is fully functional and T_{total} is the total scheduled service time. Closely related reliability indicators include the mean time between failures (MTBF), which characterizes the average operational time between successive failures:

$$\text{MTBF} = \frac{T_{\text{operational}}}{N_{\text{failures}}}, \tag{6.33}$$

where N_{failures} is the number of observed failure events.

Example 6.16: If an SLB system is scheduled to operate for 8760 hours in one year and experiences 120 hours of unplanned downtime, its availability is

$$A = \frac{8{,}760 - 120}{8{,}760} = 0.986(98.6\%).$$

If three failures occurred during this period, the MTBF is

$$\text{MTBF} = \frac{8{,}640}{3} = 2{,}880 \text{ hours.}$$

6.4.4.2 Fault Incidence and Failure Rates

Fault incidence metrics quantify the frequency and likelihood of abnormal events occurring during operation. In SLB systems, such events may arise from cell-to-cell heterogeneity, latent defects from first life, or accelerated degradation under second-life duty cycles. Common fault categories include over-voltage, over-current, abnormal temperature rise, unexpected impedance growth, and communication or sensor failures [38].

A commonly used reliability metric is the failure rate λ, defined as

$$\lambda = \frac{N_{\text{failures}}}{T_{\text{operational}}}, \tag{6.34}$$

which is often expressed in failures per 1,000 operating hours. The probability of failure over a second-life duration t can be approximated, assuming a constant failure rate, as

$$P_{\text{fail}}(t) = 1 - e^{-\lambda t}. \tag{6.35}$$

Example 6.17: If an SLB system experiences four reportable faults over 10,000 hours of operation, the failure rate is

$$\lambda = \frac{4}{10{,}000} = 4 \times 10^{-4} \text{ h}^{-1},$$

or 0.4 failures per 1000 hours. Over a planned second-life duration of 5000 hours, the corresponding probability of at least one failure is

$$P_{\text{fail}}(5{,}000) = 1 - e^{-0.0004 \times 5{,}000} = 0.86.$$

6.4.4.3 Degradation Rate and Remaining Useful Life

Quantifying degradation under second-life operating conditions is essential for predictive maintenance, service planning, and economic forecasting. Degradation metrics typically track capacity fade and internal resistance growth as functions of time, cycles, and operating stress factors such as DoD and temperature [23].

Capacity fade can be expressed as a degradation rate:

$$\dot{Q} = \frac{Q_0 - Q(t)}{t}, \tag{6.36}$$

where Q_0 is the initial usable capacity at the start of the second life and $Q(t)$ is the capacity after time t. RUL is then estimated as the time or number of cycles until a predefined end-of-life threshold Q_{EoL} is reached:

$$\mathrm{RUL} = \frac{Q(t) - Q_{\mathrm{EoL}}}{\dot{Q}}. \tag{6.37}$$

Example 6.18: Assume an SLB module enters second life with a usable capacity of 80 kWh and is considered end-of-life at 64 kWh (80% of second-life initial capacity). If capacity decreases linearly at a rate of 2 kWh per year, the RUL is

$$\mathrm{RUL} = \frac{80 - 64}{2} = 8 \text{ years.}$$

Such estimates enable scheduling of inspections, maintenance, and replacement, and directly inform availability and lifetime cost projections. Overall, reliability and operational performance metrics provide the link between cell-level degradation behavior and system-level service delivery. For SLBs, robust availability, controlled fault incidence, and well-characterized RUL are prerequisites for dependable operation and bankable techno-economic assessments.

6.4.5 Application-specific Performance Metrics

SLB suitability is inherently application dependent, as different use cases impose distinct operational, reliability, and economic constraints. Application-specific performance metrics, therefore, evaluate whether repurposed batteries can meet the functional requirements of their intended operating context without excessive degradation, loss of reliability, or economic penalty [4, 8].

6.4.5.1 Stationary Energy Storage Systems

In grid-scale and behind-the-meter stationary energy storage systems, SLBs are primarily evaluated based on their ability to deliver sustained energy throughput over the target lifetime while maintaining acceptable efficiency and degradation rates. A central metric is the cumulative energy throughput achievable before reaching end of life:

$$E_{\mathrm{throughput}} = N_{\mathrm{cycles}} \cdot \mathrm{DoD} \cdot E_{\mathrm{nom}}, \tag{6.38}$$

where N_{cycles} is the number of achievable cycles in second life, DoD is the average depth-of-discharge, and E_{nom} is the nominal energy capacity. Dynamic performance is captured through response time, typically defined as the delay between a dispatch command and achievement of the requested power level, and round-trip efficiency η_{rt}, which determines energy losses under frequent charge–discharge fluctuations [34].

Example 6.19: Consider an SLB ESS with $E_{nom} = 500$ kWh, operating at 60% DoD for 3000 cycles. The cumulative energy throughput is

$$E_{throughput} = 3000 \times 0.6 \times 500 = 900 \text{ MWh}.$$

If the system operates at a round-trip efficiency of 85%, approximately 135 MWh of energy is lost over its second-life operation.

6.4.5.2 Backup Power and Uninterruptible Power Supply

Backup power and UPS applications emphasize readiness, instantaneous power delivery, and reliability rather than continuous cycling. A key metric is the power-to-energy ratio, which determines the maximum instantaneous power that can be delivered relative to installed capacity:

$$\rho_{P/E} = \frac{P_{max}}{E_{inst}}, \tag{6.39}$$

where P_{max} is the maximum deliverable power and E_{inst} is the installed energy capacity. Operational readiness is further influenced by self-discharge rate and health drift during prolonged idle periods, which affect the guaranteed autonomy time available during outages [38].

Example 6.20: A 100 kWh SLB UPS system capable of delivering 200 kW has a power-to-energy ratio of

$$\rho_{P/E} = \frac{200}{100} = 2 \text{ h}^{-1},$$

corresponding to a 30-minute autonomy at full load. If self-discharge reduces usable capacity by 5% per month, the available autonomy must be periodically verified through maintenance cycling.

6.4.5.3 Microgrid and Off-grid Applications

In microgrid and off-grid environments, SLBs must support variable load profiles, high renewable penetration, and extended autonomy during low-generation periods. Performance is therefore assessed based on load-following capability, daily cycling endurance, and energy sufficiency:

$$T_{autonomy} = \frac{E_{usable}}{\overline{P}_{load}}, \tag{6.40}$$

where E_{usable} is the usable stored energy and $\overline{P}_{load}$ is the average load demand. Environmental robustness is also critical, as temperature extremes and high humidity can accelerate degradation and reduce effective capacity [23].

Example 6.21: If an off-grid microgrid has an average load of 15 kW and an SLB system with 300 kWh of usable energy, the autonomy time is

$$T_{\text{autonomy}} = \frac{300}{15} = 20 \text{ hours}.$$

This duration determines whether auxiliary generation or load shedding is required during extended periods of low renewable output.

6.4.5.4 Renewable Integration and Peak Shaving

For renewable integration and commercial peak shaving, SLBs are evaluated on their ability to shift energy in time and reduce peak demand charges. A key metric is peak shaving effectiveness, which quantifies the reduction in maximum demand:

$$\Delta P_{\text{peak}} = P_{\text{baseline}} - P_{\text{with SLB}}, \tag{6.41}$$

where P_{baseline} is the peak demand without storage and $P_{\text{with SLB}}$ is the peak demand after battery dispatch. Energy shifting capability is further characterized by the fraction of renewable generation shifted from low- to high-value periods, while degradation under intermittent and irregular charging profiles determines long-term viability [4].

Example 6.22: If a commercial facility reduces its monthly peak demand from 500 kW to 420 kW using an SLB system, the peak reduction is

$$\Delta P_{\text{peak}} = 80 \text{ kW}.$$

At a demand charge of 15 \$/kW·month, this corresponds to monthly savings of 1200 \$.

Remark: The metrics introduced in this subsection define what to measure for application suitability. The deployment discussion in Section 6.7 addresses where these metrics matter across grid-tied, behind-the-meter, charging-infrastructure, microgrid, and UPS deployments.

6.4.6 Multicriteria Evaluation and Composite Index Development

Because SLB performance spans technical condition, economic viability, environmental impact, operational reliability, and application-specific constraints, deployment decisions cannot be based on any single metric in isolation. Improvements in one dimension often introduce trade-offs in others; for example, aggressive DoD improves short-term economic returns but accelerates degradation, while conservative operation enhances safety and reliability at the expense of usable energy. A structured multicriteria decision-making framework is therefore required to integrate heterogeneous performance indicators into a consistent and transparent evaluation process.

Multi-criteria decision analysis (MCDA) provides such a framework by enabling disparate metrics expressed in different units, scales, and preference directions to be normalized, weighted, and aggregated into a unified decision variable. In the context of SLBs, MCDA supports systematic comparison of candidate batteries, refurbishment strategies, system configurations, and application scenarios, thereby translating raw performance metrics into actionable deployment guidance.

6.4.6.1 Metric Normalization

Prior to aggregation, all performance metrics must be mapped onto a common, dimensionless scale. This step is essential because SLB evaluation involves metrics with heterogeneous units and interpretations, including quantities that should be maximized for round-trip efficiency, availability, and RUL and others that should minimize LCOS, fault rate, and temperature rise.

For a metric x_i evaluated across alternatives j, min–max normalization is commonly applied:

$$\tilde{x}_{i,j} = \begin{cases} \dfrac{x_{i,j} - x_i^{\min}}{x_i^{\max} - x_i^{\min}}, & \text{benefit-type metric,} \\[2ex] \dfrac{x_i^{\max} - x_{i,j}}{x_i^{\max} - x_i^{\min}}, & \text{cost-type metric,} \end{cases} \tag{6.42}$$

where $x_i^{\min}$ and $x_i^{\max}$ denote the minimum and maximum observed values of the metric i among all candidates. After normalization, all metrics lie in the interval $[0,1]$, with larger values indicating more desirable performance.

Alternative mappings, such as z-score standardization or utility-based transformations, may be employed when metric distributions are highly skewed or when explicit threshold behavior is required. Regardless of the method, the objective is to preserve ordinal preference while enabling aggregation across categories.

Example 6.23: Consider two SLB candidates evaluated using LCOS, a cost-type metric. If LCOS ranges from 0.12 to 0.20 \$/kWh, and a given battery exhibits LCOS = 0.15 \$/kWh, its normalized score is

$$\tilde{x}_{\text{LCOS}} = \frac{0.20 - 0.15}{0.20 - 0.12} = 0.625,$$

indicating moderate economic attractiveness relative to the candidate set.

6.4.6.2 Weighting Strategies

Normalized metrics must be weighted to reflect their relative importance in a given deployment context. Weighting factors w_i encode stakeholder priorities, regulatory constraints, and application-specific risk tolerance. For example, grid services subject to contractual penalties may prioritize availability and fault rates, whereas behind-the-meter systems may emphasize economic metrics such as LCOS and payback time.

Weights are typically defined such that

$$\sum_{i=1}^{N} w_i = 1, \tag{6.43}$$

Ensuring the interpretability of aggregated scores. Weight selection may be based on expert judgment, policy objectives, or data-driven optimization, and should be explicitly documented to preserve transparency.

Example 6.24: For a grid-support ESS application, a decision-maker may assign weights of $w_{\text{reliability}} = 0.35$, $w_{\text{economics}} = 0.30$, $w_{\text{environment}} = 0.20$, and $w_{\text{technical}} = 0.15$, reflecting the high cost of service unavailability relative to other considerations.

6.4.6.3 Multicriteria Aggregation Methods

Once metrics are normalized and weighted, MCDA methods are used to aggregate them into scalar scores for comparison. The most straightforward approach is the weighted sum model (WSM), in which the composite score for alternative j is

$$S_j = \sum_{i=1}^{N} w_i \tilde{x}_{i,j}. \tag{6.44}$$

WSM is transparent, computationally efficient, and well suited to sensitivity analysis, making it attractive for early-stage SLB screening.

More structured methods may be employed when trade-offs are strong or qualitative judgments are significant. The analytic hierarchy process (AHP) derives weights through pairwise comparisons and consistency checks, enabling systematic incorporation of expert knowledge. The technique for order preference by similarity to ideal solution (TOPSIS) ranks alternatives based on their relative distance to an ideal best and worst solution in the normalized metric space, explicitly capturing compromise solutions. Fuzzy extensions of these methods allow uncertainty in metrics or weights to be represented, which is particularly relevant for SLBs with incomplete historical data.

Example 6.25: If three SLB candidates yield composite WSM scores of $S_1 = 0.72$, $S_2 = 0.65$, and $S_3 = 0.81$, the third candidate is ranked highest under the assumed weighting and normalization, subject to further robustness checks.

6.4.6.4 Composite Relevance Index for Second-life Deployment

The aggregated MCDA score can be interpreted as a composite relevance index that quantifies the overall suitability of a battery or system configuration for a specific second-life application. This index enables consistent ranking of candidate assets, comparison of application scenarios such as ESS, UPS, or microgrid deployment, and preliminary screening prior to detailed techno-economic or safety analysis. By construction, the composite index integrates technical condition, economic performance, environmental benefit, and operational reliability into a single decision variable, while retaining traceability to the underlying metrics through weights and normalized scores.

6.4.6.5 Sensitivity and Robustness Analysis

Given uncertainties in degradation trajectories, future electricity prices, refurbishment costs, and stakeholder priorities, robustness assessment is a critical component of multicriteria evaluation. Sensitivity analysis examines how rankings change in response to variations in metric weights, operating assumptions, or the inclusion of specific indicators. A deployment decision can be considered robust if the relative ranking of preferred alternatives remains stable under reasonable perturbations of assumptions. Sensitivity analysis, therefore, provides confidence that selected second-life pathways remain viable across plausible future conditions, reinforcing the credibility of MCDA-based deployment decisions.

6.5 Assessment and Testing of Retired EV Batteries

Before retired EV batteries can be safely and economically deployed in second-life applications, they must undergo a structured assessment and testing process. Unlike new batteries, retired packs exhibit nonuniform, path-dependent degradation shaped by first-life operating conditions, thermal exposure, and usage history, often with incomplete or unavailable data. As a result, second-life assessment must integrate diagnostics, ageing models, safety verification, and application-aware grading. This section presents a comprehensive assessment framework covering:

- battery SOH estimation and ageing models,
- use of first-life degradation data,
- screening, sorting, and repurposing strategies,
- life-cycle-oriented testing approaches,
- safety and regulatory compliance testing, and
- grading and classification for second-life applications.

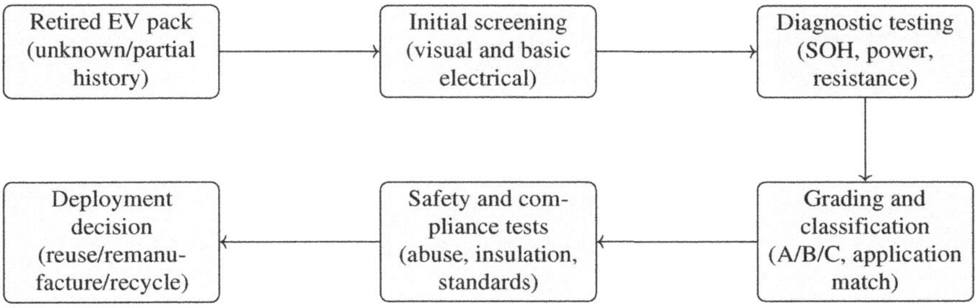

Figure 6.5 High-level workflow for assessment and testing of retired EV batteries before second-life deployment.

Together, these elements form the technical foundation for reliable techno-economic evaluation and deployment decisions. Figure 6.5 summarizes a typical workflow from battery retirement to second-life deployment.

6.5.1 Second-life Battery SOH Estimation and Aging Models

As introduced in Section 6.4.1, battery SOH reflects degradation in both energy and power capability relative to beginning-of-life (BOL) values. In the context of second-life deployment, the central challenge is not redefining SOH metrics, but accurately estimating them under heterogeneous first-life histories and projecting their evolution under new operating conditions. For retired EV batteries, SOH cannot be measured directly and must be inferred from electrical and thermal signals such as current, voltage, temperature, and impedance [22, 39]. Estimation accuracy depends on diagnostic protocol design, signal quality, and the representativeness of the applied models relative to the battery's degradation trajectory. In second-life assessment, estimation uncertainty directly affects grading decisions, system derating, warranty design, and techno-economic feasibility.

Several persistent challenges complicate SOH inference in repurposed batteries. First, ageing is path-dependent and influenced by temperature exposure, DoD distribution, C-rate severity, and storage duration. Two batteries with similar present-day capacity may exhibit markedly different internal degradation mechanisms and future degradation rates. Second, degradation mechanisms, including SEI growth, lithium plating, electrolyte decomposition, and loss of active material interact nonlinearly, making extrapolation of RUL sensitive to modeling assumptions [22, 23]. Third, available first-life operational data may be incomplete or unavailable, necessitating reliance on rapid diagnostic screening methods. SOH estimation approaches for SLBs can be broadly grouped into model-based, data-driven, and hybrid categories (Figure 6.6) [40].

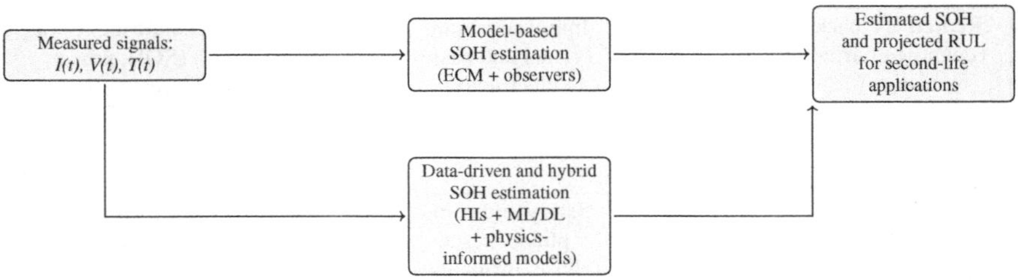

Figure 6.6 Model-based, data-driven, and hybrid SOH estimation paradigms for SLB evaluation.

Model-based techniques use equivalent circuit models (ECMs) or reduced-order electrochemical representations combined with state observers such as Kalman filtering. Aging is inferred from changes in model parameters such as ohmic resistance, polarization resistance, or diffusion-related time constants. Because these parameters have physical interpretation, model-based methods support traceability and are well aligned with certification and compliance requirements. In second-life contexts, ECM-derived parameters may be coupled with stress-factor reliability models to project degradation under stationary duty cycles [8].

Data-driven approaches use statistical learning, machine learning (ML), or deep learning (DL) algorithms to map measured signals or derived health indicators (HIs) to SOH estimates. HIs may be extracted from incremental capacity analysis (ICA), differential voltage analysis (DVA), impedance spectra, or voltage relaxation dynamics [22, 41]. Multi-indicator feature sets are particularly relevant for second-life screening, where heterogeneous degradation modes may coexist across cells drawn from different fleets.

Hybrid methods integrate physics-informed constraints within data-driven architectures. By embedding structural relationships derived from ECMs or degradation physics, hybrid frameworks can reduce data requirements while retaining the flexibility to capture nonlinear behavior. This is advantageous when historical usage data are incomplete yet limited diagnostic measurements are available.

For aging projection, semi-empirical formulations are often used to separate calendar and cycling contributions under assumed operating profiles:

$$\Delta Q(t,N) = k_{\mathrm{cal}}\sqrt{t} + k_{\mathrm{cyc}}N, \tag{6.45}$$

where t denotes storage time, N represents equivalent full cycles, and k_{cal}, k_{cyc} are stress-dependent coefficients. While simplified, such expressions provide first-order estimates for evaluating how a battery that has reached a given SOH may evolve under modified second-life duty cycles (e.g., reduced C-rate but extended calendar exposure). For rigorous applications, temperature-dependent acceleration factors or Arrhenius-type relationships may be incorporated.

Example 6.26: *Multi-Metric SOH Screening for Second-Life Suitability*
Consider a retired EV module evaluated using a standardized diagnostic protocol. The module exhibits:

- Capacity retention of 72% relative to BOL
- A 60% increase in DC internal resistance compared to BOL
- No abnormal impedance growth in mid-frequency electrochemical impedance spectroscopy (EIS).

Although the energy-related SOH exceeds a typical 70% reuse threshold, the substantial resistance growth reduces peak power capability and increases thermal losses. For a grid-tied peak-shaving application with low C-rate requirements, the module may remain acceptable under conservative derating. However, for applications requiring rapid frequency response, the elevated resistance may render it unsuitable. This example illustrates that second-life screening must interpret SOH metrics in conjunction with application-specific power and thermal constraints rather than relying on a single numerical threshold.

Table 6.3 summarizes representative SOH estimation approaches and their practical trade-offs in second-life assessment.

The second-life SOH estimation extends beyond calculating degradation metrics; it requires integrating diagnostics, aging modeling, uncertainty quantification, and application-aware interpretation. Multidimensional evaluation is therefore essential to support reliable grading, system integration decisions, and credible lifetime projection in repurposed battery systems.

Table 6.3 Representative SOH estimation approaches for retired EV batteries.

Method	Key inputs	Advantages	Limitations
Coulomb counting [39]	$I(t)$, time	Simple, low computational cost	Drift accumulation; requires reference calibration
ECM + observers [22]	$V(t), I(t)$, ECM parameters	Physics-informed, interpretable	Parameter sensitivity; model mismatch risk
HI-based ML [40]	Voltage, impedance, ICA features	Captures nonlinear ageing behavior	Data dependency; limited extrapolation robustness
Hybrid physics-ML	Multi-domain signals	Improved generalization under limited data	Higher implementation complexity

6.5.2 Degradation Data from First Use

Access to first-life operational data significantly improves the accuracy and confidence of SLB assessment. Unlike laboratory characterization alone, historical usage records provide direct evidence of cumulative electrical, thermal, and mechanical stress experienced during automotive service. Modern BMS, particularly cloud-connected architectures, increasingly log detailed information including current profiles, voltage trajectories, temperature distributions, depth-of-discharge (DOD) statistics, and fault events [8, 22]. When available and reliable, these datasets enable reconstruction of degradation pathways and reduce uncertainty in residual performance estimation.

From a degradation modeling perspective, battery aging is commonly decomposed into calendar and cycling contributions. The cumulative degradation can be expressed as

$$\Delta Q(t,N) = \Delta Q_{\text{cal}}(t,T) + \Delta Q_{\text{cyc}}(N,\text{DOD},I), \tag{6.46}$$

where t denotes the time in service, N represents the equivalent full cycles, T is the temperature, and I is the current magnitude. Calendar aging is frequently modeled using a square-root time dependence:

$$\Delta Q_{\text{cal}} = k_{\text{cal}}(T,\text{SoC})\,\sqrt{t}, \tag{6.47}$$

while cycle aging is often approximated as proportional to accumulated charge throughput:

$$\Delta Q_{\text{cyc}} = k_{\text{cyc}}(T,\text{DOD}) \cdot Q_{\text{throughput}}. \tag{6.48}$$

Historical BMS data allow estimation of stress factors such as high C-rate exposure, prolonged operation at elevated temperature, and frequent deep DOD cycling. For example, temperature acceleration effects are commonly described using Arrhenius-type relationships:

$$k(T) = k_0 \exp\left(-\frac{E_a}{RT}\right), \tag{6.49}$$

where E_a is activation energy, R is the universal gas constant, and T is absolute temperature. Thus, modules exposed to higher average temperatures may exhibit disproportionately accelerated degradation compared to modules operating under moderate conditions.

Cloud-connected BMS architectures facilitate aggregation of large-scale fleet data, enabling empirical aging models that extend predictions across both first and second life [40]. Such frameworks can be particularly valuable when assessing batteries with incomplete physical testing, provided chemistry consistency and data quality are maintained. Nevertheless, heterogeneity within retired packs remains a central challenge. Cells within the same pack may have experienced uneven thermal gradients, current distribution imbalance, or localized degradation, leading to dispersion in capacity and resistance [23]. Consequently, first-life data must be interpreted at both pack and module levels.

Example 6.27: *First-Life Data–Informed Screening*

Consider two retired EV modules, A and B, each originally rated at 50 Ah. Historical BMS records indicate the following:

- Module A: 1200 equivalent full cycles at 25 °C average temperature, average DOD = 60%.
- Module B: 1200 equivalent full cycles at 40 °C average temperature, average DOD = 90%.

Assume a simplified degradation model:

$$\Delta Q = k_{\mathrm{cal}}\sqrt{t} + k_{\mathrm{cyc}} \cdot N_{\mathrm{eq}},$$

where k_{cal} doubles for every 10 °C temperature increase (Arrhenius approximation), and k_{cyc} scales linearly with DOD.

If Module A exhibits a measured capacity of 38 Ah after first life, its capacity-based SOH is:

$$\mathrm{SOH}_A = \frac{38}{50} \times 100\% = 76\%.$$

Due to elevated temperature and deeper cycling, Module B experiences accelerated degradation and measures 32 Ah:

$$\mathrm{SOH}_B = \frac{32}{50} \times 100\% = 64\%.$$

Although both modules underwent the same nominal cycle count, first-life stress exposure results in markedly different residual capacity. In a second-life screening context, Module A may qualify for medium-power stationary storage, whereas Module B may require derating or redirection to lower-demand applications. This example illustrates how first-life data directly informs screening thresholds, grading decisions, and repurposing strategy. Without historical information, both modules might initially appear equivalent based solely on cycle count, leading to suboptimal classification.

It is noticed that the incorporation of first-life degradation data enhances predictive accuracy, reduces testing uncertainty, and enables application-aware grading of retired EV batteries. However, data incompleteness, measurement noise, and intra-pack heterogeneity necessitate complementary electrical and thermal diagnostics before final second-life deployment decisions are made. These insights motivate screening and sorting procedures that explicitly account for dispersion and uncertainty when regrouping assets for second-life assembly.

6.5.3 Screening, Sorting, and Repurposing

Screening and sorting represent the critical transition from diagnostic evaluation to practical second-life deployment [14]. While SOH estimation quantifies residual capability, screening determines whether a battery module is suitable for continued use, and sorting defines how heterogeneity among retired assets is managed. Because retired EV batteries exhibit dispersion in capacity, resistance, and thermal behavior, systematic classification and grouping are essential to ensure safe and reliable repurposing.

A typical screening workflow includes initial safety inspection, electrical characterization, including capacity and resistance measurement, consistency evaluation across cells or modules, sorting based on predefined thresholds, and finally application-specific validation [42]. Experimental investigations consistently demonstrate that second-life performance strongly depends on grouping strategy and operational constraints. In particular, operation at reduced C-rates significantly mitigates stress and prolongs usable life in stationary applications. Impedance-informed grouping further improves pack-level homogeneity by aligning modules with similar electrochemical characteristics, thereby reducing imbalance and localized overstress.

Beyond simple threshold-based sorting, advanced batching and energy management strategies have been proposed to address dispersion effects in heterogeneous assemblies. Such strategies may include dynamic load allocation, hierarchical battery management, and modular reconfiguration, allowing weaker units to operate under less demanding conditions. These approaches are especially relevant for large-scale battery energy storage systems (BESS), where economic viability depends on maximizing utilization of available assets while preserving safety margins [43].

6.5.3.1 Advanced Repurposing and Testing Techniques

In addition to conventional charge–discharge characterization, several advanced repurposing methodologies have been developed to enhance reliability and extend second-life viability [44]. The multi-level interlaced pulse charging (MLIPC) technique combines pulsed charge management with battery interlacing. By using multiple interlaced PWM pulse trains, this method promotes controlled charge redistribution and reduces localized stress concentration on individual cells. Although promising from a technical standpoint, further techno-economic analysis and parameter optimization are required before large-scale deployment [44].

For batteries with incomplete or unknown usage history, multi-method diagnostic protocols offer improved robustness, such as hybrid pulse power characterization (HPPC), electrochemical impedance spectroscopy (EIS), and incremental capacity analysis (ICA). Each diagnostic modality captures different degradation signatures: HPPC assesses dynamic power capability, EIS reveals impedance growth and electrochemical shifts, and ICA highlights structural changes through differential capacity features. The combined use of these methods provides a more comprehensive basis for sorting compared to any single diagnostic technique [45].

Automation is also emerging as a key enabler for scalable second-life operations. The automated testing platform is capable of performing accelerated life tests and generating real-time feasibility data for large volumes of retired LIBs. Such systems reduce manual intervention, improve repeatability, and enable data-driven grading at the industrial scale. As second-life markets expand, automated characterization and digitalized testing infrastructures will become increasingly important for economic competitiveness. Table 6.4 summarizes representative experimental approaches and their implications for screening and repurposing strategies.

6.5.4 Safety and Regulatory Compliance Testing

Safety verification is a mandatory prerequisite for second-life deployment of retired EV batteries. Unlike performance-oriented diagnostics, safety assessment focuses on identifying latent defects, degradation-induced vulnerabilities, and failure risks that may compromise stationary operation. The evaluation typically begins

Table 6.4 Representative screening and repurposing approaches for retired EV batteries.

Study	Main techniques	Key insight for second-life
Accelerated Test	Accelerated cyclic testing, capacity and resistance tracking	Identification of aging knee and low-performance second-life suitability
SOH-based	Performance tests at various SOH levels	Low-rate operation extends second life in frequency services
EIS-based	EIS, impedance circuit parametrization	Chemically-informed SOH estimation and degradation mapping
Smart batching	Capacity tests, batching algorithms	Reduced dispersion and improved pack homogeneity
HPPC + EIS + ICA	Multi-method diagnostics	Robust sorting for unknown-history cells

with nondestructive inspection, including visual examination for mechanical damage, corrosion, electrolyte leakage, and connector integrity, followed by insulation resistance testing and verification of electrical isolation.

Thermal behavior and abuse tolerance are particularly critical in stationary energy storage systems, where batteries may operate for extended periods under varying environmental conditions. Controlled abuse testing such as overcharge, over-discharge, short-circuit, and elevated-temperature exposure is conducted to evaluate stability margins and failure propagation characteristics. Special attention is given to thermal runaway initiation and propagation risk, especially in heterogeneous second-life assemblies where degradation dispersion may increase vulnerability.

Compliance with established safety and repurposing standards is essential to ensure regulatory acceptance and market readiness. In North America, UL 1973 governs stationary battery systems, while UL 1974 specifically addresses evaluation and grading procedures for repurposed batteries. Internationally, IEC 62619 specifies safety requirements for secondary lithium batteries in industrial applications. These standards define testing protocols, performance thresholds, documentation requirements, and traceability procedures necessary for certification.

In the context of SLBs, regulatory compliance not only ensures safe integration into stationary systems but also provides a structured framework for risk mitigation, warranty definition, and liability management. As second-life markets mature, harmonization of testing methodologies and certification pathways will play a central role in enabling large-scale deployment.

6.6 Grading, Classification, and Application Matching

Following assessment and testing (Section 6.5), retired EV batteries must be translated into a clear deployment decision. This transition is achieved through a structured grading and classification stage that functions as a decision gate prior to second-life integration. The purpose of grading is threefold: (i) to reduce uncertainty and safety risk by enforcing minimum technical thresholds, (ii) to manage heterogeneity through systematic categorization, and (iii) to align residual battery capability with application-specific power, energy, and reliability requirements. In practice, grading operationalizes the metrics of Section 6.4 by converting capacity, resistance, thermal headroom, and reliability indicators into enforceable operating envelopes (SoC window, current limits, and derating rules) suitable for specific applications.

6.6.1 Purpose and Decision-gate Role

Grading converts multi-dimensional diagnostic outcomes such as capacity-based SOH, resistance growth, thermal response, self-discharge behavior, and intra-pack dispersion into operationally actionable classes. These classes can then be directly embedded into refurbishment workflows, warranty design, and techno-economic evaluation models. In practice, grading supports three principal outcomes: (i) direct reuse under defined derating and monitoring policies, (ii) conditional reuse for low-demand or constrained applications, and (iii) recycling or remanufacturing when technical or safety thresholds are not satisfied. By acting as a structured filtering mechanism, grading reduces the risk of premature failure in second-life systems and establishes a transparent basis for value estimation.

6.6.2 Grade A/B/C Logic and Thresholding

A practical grading scheme typically classifies batteries into three categories based primarily on usable capacity (or capacity-based SOH) and internal resistance growth, complemented by safety verification outcomes. Grade A batteries generally exhibit usable capacity above 80% with moderate resistance increase and are suitable for applications involving regular cycling and moderate power demand. Grade B batteries, typically within the 60–80% capacity range, exhibit higher resistance and are better suited for lower-power or less intensive duty cycles. Grade C batteries fall below 60% capacity and/or exhibit excessive resistance growth; these assets are often restricted to niche low-demand use or directed toward recycling.

In addition to these quantitative indicators, several hard gating criteria are commonly applied. Batteries must pass insulation resistance and dielectric withstand tests, demonstrate acceptable leakage current, and self-discharge behavior, and satisfy defined abuse-test thresholds where required. Thermal performance under representative load profiles is also critical; excessive temperature rise or unstable thermal behavior may trigger derating or rejection. Finally, dispersion constraints at the cell or module level must be considered, as excessive heterogeneity increases balancing losses and long-term reliability risk.

6.6.3 Capacity–Resistance Trade-offs and Power Capability

Capacity- and resistance-based SOH indicators often diverge for the same retired battery (Section 6.5.1), underscoring the need for multi-indicator grading. A battery with acceptable residual capacity but elevated resistance may retain substantial energy value while suffering degraded power capability. Increased resistance leads to reduced peak current limits, higher I^2R losses, lower round-trip efficiency, and increased thermal stress. Consequently, capacity primarily governs usable energy, whereas resistance governs power value and thermal risk. Robust grading frameworks, therefore, treat the capacity–resistance pair as a joint decision variable rather than relying on a single SOH metric.

Example 6.28: A simple threshold-based grading rule may define Grade A if $SOH \geq 80\%$ and $R/R_{new} \leq 1.5$, Grade B if $60\% \leq SOH < 80\%$ and $R/R_{new} \leq 2.0$, and Grade C otherwise. Table 6.5 presents an illustrative classification outcome.

Under this scheme, M1 may be assigned to higher-demand cycling applications, M2 to moderate-duty stationary use, and M3 either to highly constrained service or recycling.

6.6.4 Mapping Grades to Second-life Applications

Once grades are established, batteries must be mapped to applications using an application-requirement matrix that reflects power intensity, cycling frequency, and reliability criticality. Table 6.6 provides an illustrative mapping between grade classification and second-life application categories. This mapping can be extended to incorporate explicit quantitative requirements—such as minimum

Table 6.5 Example grading of retired EV battery modules.

Module	Capacity SOH	R/R_{new}	Grade
M1	85%	1.3	A
M2	72%	1.6	B
M3	58%	2.1	C

Table 6.6 Illustrative mapping from battery grades to second-life application classes.

Grade	Best-fit applications	Notes/constraints
Grade A	C&I PV-firming, residential ESS, microgrid cycling, moderate-frequency services	Higher cycling tolerance; requires defined derating and monitoring
Grade B	Backup/UPS, telecom backup, peak shaving with limited DoD, low-to-moderate cycling ESS	Power-limited; operate at lower C-rate and tighter thermal margins
Grade C	Emergency lighting, very low-demand backup, pilot deployments, or recycling	Strict constraints; often the lowest economic value

usable capacity, maximum allowable resistance, maximum temperature rise, and target RUL, enabling automated assignment in industrial-scale repurposing operations.

6.6.5 Interface with Techno-economic Assessment

Grading outputs serve as direct inputs to techno-economic analysis by constraining feasible operating windows and expected service life. Usable capacity and efficiency determine lifetime energy throughput and influence LCOS or LCOE calculations, while resistance-driven power limits shape revenue potential in power-intensive services. Grade-dependent reliability risk affects maintenance cost assumptions, warranty structure, and discount factors, and predicted RUL determines reinvestment or replacement schedules. Grading therefore represents the critical interface between physical battery condition and economic feasibility. By linking diagnostic outcomes to operational strategy and financial modeling, the grading framework enables structured, defensible second-life deployment decisions.

6.7 Deployment Domains and Case Studies for Second-life Batteries

This section translates the performance metrics (Section 6.4) and the assessment and grading framework (Sections 6.5–6.6.5) into practical deployment contexts. The objective is to identify where SLB energy storage systems (SLBESSs) create technical and economic value, and to clarify how application requirements map to operating windows (SoC and DoD constraints), power-electronics limitations (power rating, ramp rate, latency), thermal margins, and revenue mechanisms (tariffs, ancillary markets, avoided infrastructure upgrades). Second-life feasibility is inherently application-dependent. Energy-intensive services emphasize usable capacity and round-trip efficiency, while power-intensive services are constrained by resistance growth, inverter capability, and thermal headroom. Proper alignment between residual capability and duty cycle is therefore essential to maximize lifetime and economic return.

6.7.1 Stationary Second-life Applications

Stationary deployment represents the most mature and scalable pathway for SLBs. Unlike automotive applications, stationary systems relax weight and volume constraints, enable controlled thermal environments, and permit structured integration with inverters and supervisory energy management systems. In these settings, requirements are typically governed by usable capacity, operational availability, and moderate power responsiveness rather than extreme power density. Accordingly, SLBESSs are particularly suited to grid-connected flexibility services, behind-the-meter (BTM) energy management, charging-infrastructure buffering, and microgrid or UPS support.

6.7.1.1 Grid-tied Energy Storage

6.7.1.1.1 Use Cases

Grid-tied SLBESSs are deployed for peak shaving, demand-charge reduction, energy time-shifting (arbitrage), frequency regulation, voltage support, and transmission/distribution (T&D) deferral. These services rely on controllable real-power injection or absorption through power-electronic interfaces and dispatch strategies aligned with local tariffs and market structures. As renewable penetration and electrified loads increase, grid operators require flexible assets to mitigate short-term imbalances, reduce congestion, and defer infrastructure upgrades. In this context, SLBESSs can provide technically feasible and economically attractive flexibility, provided degraded power capability and safety margins are explicitly incorporated into dispatch policies and derating rules (Figure 6.7).

6.7.1.1.2 Representative Grid Services

Peak shaving and energy arbitrage are primarily energy-driven services. The SLBESS charges during low-demand or low-price periods and discharges during peak periods to reduce feeder loading and demand charges. These services are generally compatible with Grades A and B systems operated at moderate C-rates within conservative SoC windows. Frequency services and short-term balancing are power-driven. They require high ramp rates, low effective DCIR, sufficient inverter rating, and robust thermal headroom. Here, resistance growth and transient thermal response become dominant constraints. T&D deferral relies on predictable discharge during peak loading windows to reduce apparent transformer or feeder stress. This application

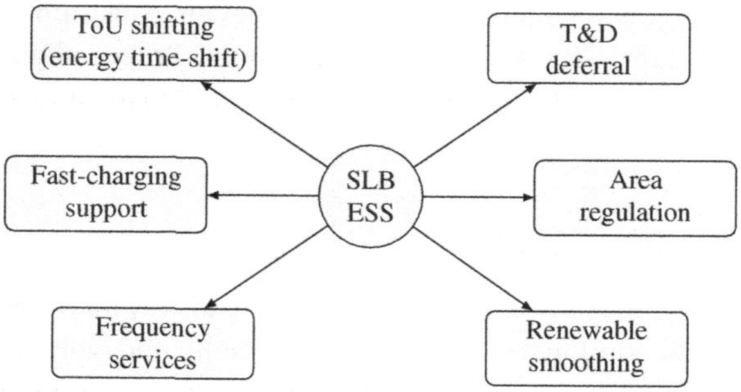

Figure 6.7 Ecosystem of grid-flexibility and support services enabled by SLBESSs.

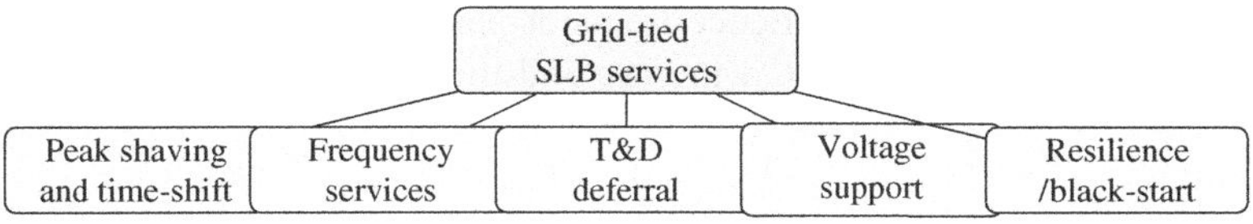

Figure 6.8 Core grid-tied service classes for stationary SLBs.

Table 6.7 Grid-tied services enabled by SLBs and typical technical requirements.

Service	Description	Typical requirements
Peak shaving/ arbitrage	Shift energy across ToU periods; reduce peak demand	Moderate power, daily cycles, adequate SOH, good RTE, conservative SoC window
Frequency services	Rapid balancing of short-term supply/demand	High ramp rate, sufficient power capability, low DCIR, thermal headroom, low latency
T&D deferral	Support constrained transformers/feeders during peaks	Predictable discharge windows, high availability, reliability, conservative derating
Voltage support	Maintain LV/MV voltage and mitigate fluctuations	Stable control, reactive capability, fast response, robust communications
Black-start/ resilience	Energize local circuits during outages	High availability, protection coordination, verified fault handling

emphasizes reliability, conservative derating, and high availability rather than continuous cycling. Voltage regulation and power-quality support depend primarily on inverter control capability and communication stability. While energy requirements are modest, dynamic control performance is critical. Black-start and resilience support are event-driven and emphasize availability, protection coordination, and safe restart capability rather than throughput energy (Figure 6.8). A summary of grid-tied services enabled by SLBs and their technical requirements are summarized in Table 6.7.

6.7.1.2 Residential and Commercial ESS

BTM SLBESS deployments generate value primarily through bill reduction, increased on-site renewable utilization, and resilience enhancement. Compared with utility-scale systems, BTM applications are strongly influenced by tariff structures (e.g., time-of-use pricing and demand charges), customer reliability requirements, and ease of integration with existing electrical infrastructure.

Operationally, these systems experience frequent partial cycling and therefore require robust management of SoC windows and thermal margins to preserve lifetime.

6.7.1.2.1 PV Self-consumption and Residential Microgrids

In residential environments, SLBESSs store excess photovoltaic (PV) generation during midday peaks and discharge during evening demand peaks, increasing self-consumption and reducing grid imports. Daily load-shifting capacities are often on the order of a few kilowatt-hours, typically 3–4 kWh, operated at relatively low C-rates with mixed cycling patterns consisting of one deeper cycle supplemented by shallow cycles.

For extended backup or off-grid configurations, larger capacities in the range of 20–25 kWh may be deployed, often within moderated DoD windows to limit degradation. In residential microgrids, temporal mismatch between PV generation and household demand reduces direct utilization; coupling PV with SLBESSs improves energy autonomy and may reduce capital cost relative to new-battery installations, particularly when refurbishment costs are favorable (Figure 6.9).

6.7.1.2.2 Commercial and Industrial (C&I) Deployments

C&I facilities often experience higher peak demand, significant demand charges, and stricter reliability constraints. SLBESSs can provide peak shaving, load following for variable industrial processes, and backup power for critical loads. Regular discharge rates may range from approximately C/3 to C under scheduled operation, while backup operation typically occurs at lower average rates. In cases where peak demand reduction targets exceed safe SLB power capability, hybrid

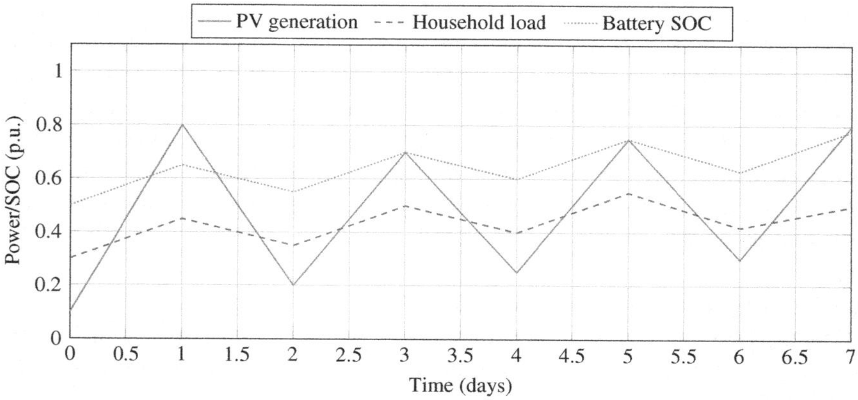

Figure 6.9 Illustrative interaction between PV generation, residential load, and SLB state of charge in a PV–SLB microgrid.

configurations combining second-life and new modules may be appropriate. Conversely, moderate load-following or resilience applications can often be supported entirely by SLBs operating within conservative SoC limits.

6.7.1.3 Second-life Batteries in EV Charging Infrastructure

Fast-charging infrastructure imposes short-duration, high-power peaks that can exceed local distribution-network capacity. Integrating an SLBESS buffer at the charging site reduces required grid interconnection size, lowers demand charges, and mitigates voltage fluctuations during simultaneous vehicle charging events. In this configuration, the SLBESS charges during low-demand periods and discharges when multiple vehicles arrive concurrently. Because this application is strongly power- and thermal-limited, conservative derating, verified protection coordination, and careful thermal management are essential. While economically attractive in constrained networks, long-term viability depends on managing resistance growth and transient thermal stress (Figure 6.10).

6.7.1.4 Microgrids, Renewables, and UPS

SLBESSs are well suited to microgrids and renewable-rich systems, particularly in remote or weak-grid regions. When integrated with PV or wind generation, they mitigate output variability, reduce ramp rates, and enhance frequency and voltage stability. These roles increase in importance as renewable penetration rises, although improper sizing or aggressive dispatch strategies may accelerate degradation and increase life-cycle cost. In off-grid systems, SLBs provide autonomy and reliability where utility supply is unavailable or unstable. In UPS applications, the emphasis shifts to short-duration ride-through capability with high availability and predictable response. In all cases, application-aligned SoC windows and moderate C-rates improve durability.

6.7.2 Illustrative Lifetime Sensitivity Across Use Cases

Second-life duration depends strongly on imposed duty cycle and operating environment. Even with identical initial SOH, distinct applications yield different degradation trajectories due to variations in throughput energy, C-rate, temperature

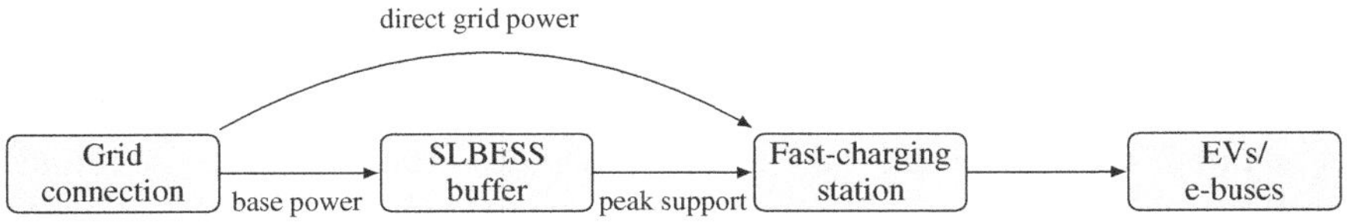

Figure 6.10 Conceptual SLB buffer between the grid and fast-charging stations to reduce peak demand.

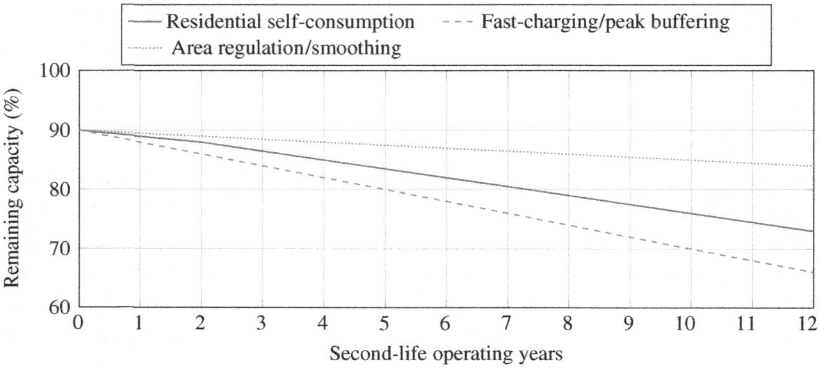

Figure 6.11 Schematic aging trajectories of SLBs under different second-life use cases.

exposure, and SoC window. Power-intensive services such as peak buffering or fast charging generally accelerate resistance growth and thermal stress, whereas moderate cycling in residential self-consumption applications may extend service life. These schematic profiles emphasize that application matching and conservative dispatch strategies are central to maximizing second-life value (Figure 6.11).

6.7.3 Economic Note on Service Stacking

Single-service deployments may yield extended payback periods in certain markets, whereas multi-service operation (service stacking) improves utilization and revenue potential. For example, a commercial SLBESS may simultaneously perform peak shaving, energy arbitrage, and resilience support, while a charging-site SLBESS may combine peak buffering with tariff optimization. Economic performance depends strongly on tariff design, ancillary-service markets, refurbishment cost, and regulatory structure. Nevertheless, combining complementary services typically enhances cash flow stability and shortens payback time relative to single-service operations. Table 6.8 presents an illustrative example of how combining multiple SLB services can affect project economics.

6.8 Future Trends and Innovations

SLB deployment is evolving from small-scale repurposing initiatives toward structured, data-informed, and system-integrated value chains. Continued progress will depend on improvements in diagnostic fidelity, uncertainty-aware degradation modeling, refurbishment standardization, and integration into regulated electricity markets. Rather than being driven solely by electrochemical

Table 6.8 Illustrative impact of combining multiple SLB services on project economics.

Combination of services	Typical payback	Notes
Fast charging + peak shaving	3–5 years	Reduces demand charges and grid capacity requirements
Frequency regulation + arbitrage	2–4 years	Revenue stacking in favorable ancillary markets
PV self-consumption + peak shaving	5–7 years	Improves behind-the-meter PV utilization
PV + fast charging + arbitrage	4–6 years	Suitable for commercial sites and fleet depots

performance, the next phase of SLB development will be shaped by advances in digital infrastructure, operational intelligence, and risk allocation frameworks.

6.8.1 Digital-twin-based Diagnostics and Health-aware Operation

Digital-twin frameworks are increasingly investigated for battery systems in both first- and second-life contexts. A digital twin maintains a continuously updated virtual representation of a battery module or system by synchronizing real-time measurements with physics-informed or hybrid degradation models. For SLBs, such frameworks can support structured refurbishment workflows and operational risk management.

In second-life deployment, digital twins can enable:

- rapid screening and grading by integrating diagnostic measurements with model-based parameter estimation,
- scenario-based RUL projection under candidate stationary duty cycles,
- detection of abnormal impedance growth or thermal deviation indicative of emerging faults,
- adaptive operating-window control that adjusts current and voltage limits according to estimated health state.

For example, a digital twin calibrated during refurbishment may be used to simulate peak-shaving duty profiles and evaluate projected degradation acceleration before commissioning. This capability allows operators to define conservative derating policies and maintenance intervals based on quantified uncertainty rather than fixed assumptions. By linking diagnostics with operational control, digital twins can reduce failure probability and support performance-backed service contracts.

6.8.2 AI-driven SOH and RUL Estimation Under Heterogeneity

Second-life batteries originate from fleets with diverse chemistries, usage histories, and thermal exposure. This heterogeneity introduces domain-shift challenges when applying degradation models developed for homogeneous first-life datasets. Artificial intelligence (AI) and machine learning (ML) techniques are therefore increasingly explored to improve robustness across heterogeneous asset pools.

Advanced approaches utilize multi-modal health indicators, including incremental capacity signatures, differential voltage features, impedance spectra, and relaxation dynamics. Rather than relying on a single degradation proxy, multi-feature models can better distinguish between capacity-limited and resistance-limited aging modes.

Key technical priorities include:

- transfer learning methods that adapt models trained on first-life data to second-life conditions,
- probabilistic SOH and RUL estimation with explicit uncertainty bounds,
- interpretable architectures compatible with safety certification and audit requirements,
- computationally efficient inference suitable for embedded BMS or edge platforms.

Hybrid physics-informed learning approaches, which constrain data-driven models using degradation physics or equivalent-circuit structure, are particularly promising for SLB applications. These methods can improve extrapolation reliability when available historical data are incomplete.

6.8.3 Standardized Refurbishment and Integration Architectures

One of the principal barriers to scalable SLB deployment is the diversity of module formats, electrical interfaces, and communication protocols across EV platforms. Without architectural harmonization, refurbishment cost, and certification complexity remain high.

Standardized refurbishment strategies may include modular rack designs with configurable voltage windows, common DC bus interfaces, unified communication layers for BMS integration, and pre-qualified thermal management subsystems. Such standardization can:

- reduce engineering effort required for mixed-grade asset integration,
- simplify safety validation and regulatory compliance,
- enable repeatable derating envelopes and warranty structures,
- facilitate traceable end-of-second-life recycling.

As regulatory oversight increases, structured refurbishment architectures are likely to become a prerequisite for large-scale commercial SLB markets rather than an optional optimization.

6.8.4 Service-based Business Models and Risk Allocation

Economic innovation will play a central role in determining SLB market penetration. Service-oriented frameworks—often described as battery-as-a-service (BaaS) separate asset ownership from performance delivery and may reduce upfront capital exposure for end users.

In second-life applications, such models can incorporate:

- grade-dependent pricing structures linked to validated SOH metrics,
- performance guarantees based on uncertainty-aware RUL projections,
- contract structures aligned with peak shaving, resilience, or flexibility services,
- predefined end-of-second-life pathways, including recycling obligations.

The effectiveness of these models depends critically on credible diagnostics and transparent degradation tracking. Financial stakeholders require quantifiable risk bounds; therefore, technical advances in health estimation directly influence capital cost and project bankability.

6.8.5 Grid-interactive and Aggregated SLB Fleets

As distributed SLB installations expand, aggregation into coordinated fleets will enable participation in grid services such as frequency regulation, congestion mitigation, and peak demand management. Virtual power plant (VPP) configurations allow geographically dispersed assets to operate as a single controllable entity.

Fleet-level coordination introduces additional technical requirements:

- secure and low-latency telemetry,
- standardized control and communication interfaces,
- dispatch algorithms compliant with market rules,
- degradation-aware optimization that accounts for heterogeneous asset health.

For instance, fleet control strategies may preferentially dispatch higher-SOH modules for high-power events while reserving lower-SOH assets for energy-shifting services. Such health-aware scheduling can extend overall fleet lifetime while maintaining revenue participation.

The future SLB systems will be defined not only by residual electrochemical capability, but by integration of advanced diagnostics, uncertainty-aware modeling, standardized refurbishment practices, service-oriented financing, and coordinated grid interaction. Ultimately, scalable second-life deployment will depend on standardized qualification protocols, uncertainty-aware lifetime forecasting, digitalized testing infrastructures, and interoperable control architectures that collectively enhance bankability and long-term safety assurance.

6.9 Conclusions and Outlook

This chapter reviewed the second-life deployment of retired EV batteries from an energy and electrical engineering perspective. We introduced a mobility-based application taxonomy, defined decision-relevant performance metrics, including technical, economic, environmental, and reliability, and described structured workflows for assessment, screening, grading, and application matching. Across use cases, deployment feasibility is governed by the coupled effects of capacity fade, resistance growth, efficiency losses, and thermal safety margins, all under significant heterogeneity and uncertainty in RUL. Future progress will be driven by improved diagnostics, uncertainty-aware lifetime prediction, standardized refurbishment architectures, and scalable testing/qualification pipelines that strengthen bankability. When combined with appropriate derating and health-aware operation, SLBs can become a practical contributor to stationary flexibility needs while complementing end-of-life recycling pathways.

6.10 Additional Solved Examples

Example 6.29: (Power Capability Constraint from Resistance Growth):

A second-life module has a nominal voltage of 400 V. *Its internal resistance has increased from* 40 mΩ *(new) to* 80 mΩ. *If the maximum allowable voltage drop under discharge is limited to 10% of nominal voltage, determine the maximum permissible discharge current.*

Solution: Maximum allowable voltage drop:

$$\Delta V_{\text{max}} = 0.10 \times 400 = 40 \text{ V}.$$

Using Ohm's law:

$$I_{\text{max}} = \frac{\Delta V_{\text{max}}}{R} = \frac{40}{0.08} = 500 \text{ A}.$$

Thus, the maximum discharge current is **500 A**. If resistance had remained at 40 mΩ, the limit would have been 1000 A, demonstrating how aging directly reduces usable power capability.

Example 6.30: (Demand-charge Reduction Economics):

A commercial facility faces a demand charge of \$18/kW-month. Its peak demand exceeds the contracted limit by 35 kW. An SLBESS is installed to shave this excess peak. Estimate the approximate annual demand-charge savings.

Solution: Monthly avoided demand cost:

$$35 \times 18 = \$630.$$

Annual savings:

$$630 \times 12 = \$7{,}560 \text{ per year.}$$

This simplified estimate excludes efficiency losses and degradation costs, but illustrates the strong economic driver of peak shaving for SLB deployment.

Example 6.31: (Levelized Storage Cost (LSC) Estimation):

A refurbished SLB system costs \$120/kWh (capital). Annual O&M cost is \$5/kWh-year. The system cycles 250 times per year, with an average usable throughput of 0.5 kWh per kWh installed per cycle. Expected second-life duration is 6 years. Estimate the levelized storage cost (LSC) in \$/kWh delivered.

Solution: Total lifetime throughput per kWh installed:

$$E_{\text{th}} = 250 \times 0.5 \times 6 = 750 \text{ kWh.}$$

Total cost per kWh installed:

$$C_{\text{tot}} = 120 + 5 \times 6 = \$150.$$

Therefore,

$$\text{LSC} = \frac{150}{750} = \$0.20/\text{kWh.}$$

This value can be compared with tariff spreads or ancillary market revenues to assess project viability.

Example 6.32: (Microgrid Ride-through Energy Requirement):

A critical load of 12 kW *must be supported for* 20 min *during grid disturbances. If the SLB system operates with allowable DoD = 70% and round-trip efficiency = 90%, what minimum nominal energy capacity (kWh) should be installed?*

Solution: Required delivered energy:

$$E_{\text{load}} = 12 \times \frac{20}{60} = 4 \text{ kWh.}$$

Accounting for efficiency and allowable DoD:

$$E_{\text{nom}} = \frac{E_{\text{load}}}{0.90 \times 0.70} = \frac{4}{0.63} \approx 6.35 \text{ kWh}.$$

A practical design would specify at least **6.5–7 kWh** to include operational margin.

References

1 J. Zhu, et al., "End-of-life or second-life options for retired electric vehicle batteries," *Cell Rep. Phys. Sci.*, vol. 2, no. 8, 2021.

2 R. Sathre, C. D. Scown, O. Kavvada, and T. P. Hendrickson, "Energy and climate effects of second-life use of electric vehicle batteries in California through 2050," *J. Power Sources*, vol. 288, pp. 82–91, 2015.

3 M. F. Börner, M. H. Frieges, B. Späth, K. Spütz, H. H. Heimes, D. U. Sauer, and W. Li, "Challenges of second-life concepts for retired electric vehicle batteries," *Cell Rep. Phys. Sci.*, vol. 3, no. 10, 2022.

4 L. C. Casals, B. A. García, F. Aguesse, and A. Iturrondobeitia, "Second life of electric vehicle batteries: Relation between materials degradation and environmental impact," *Int. J. Life Cycle Assess.*, vol. 22, no. 1, pp. 82–93, 2017.

5 I. Stefan and K. Chirumalla, "Enabling value retention in circular ecosystems for the second life of electric vehicle batteries," *Resour. Conserv. Recycl.*, vol. 212, p. 107942, 2025.

6 C. Altuntas Vural, P. van Loon, Á. Halldórsson, J. Fransson, and F. Josefsson, "Life after use: circular supply chains for second-life of electric vehicle batteries," *Prod. Plan. Control*, vol. 36, no. 9, pp. 1229–1246, 2025.

7 J. Neubauer and A. Pesaran, "The ability of battery second use strategies to impact plug-in electric vehicle prices and serve utility energy storage applications," *J. Power Sources*, vol. 196, no. 23, pp. 10 351–10 358, 2011.

8 J. S. Neubauer, E. Wood, and A. Pesaran, "A second life for electric vehicle batteries: Answering questions on battery degradation and value," *SAE Int. J. Mater. Manuf.*, vol. 8, no. 2, 2015.

9 E. Martinez-Laserna, et al., "Battery second life: Hype, hope or reality? a critical review of the state of the art," *Renew. Sustain. Energy Rev.*, vol. 93, pp. 701–718, 2018.

10 D. Svensson and A. Särnevång, "Inspection and evaluation method to repurpose used batteries," 2024.

11 R. Faria, et al., "Primary and secondary use of electric mobility batteries from a life cycle perspective," *J. Power Sources*, vol. 262, pp. 169–177, 2014.

12 J. Neubauer, A. Pesaran, B. Williams, M. Ferry, and J. Eyer, "Techno-economic analysis of pev battery second use: Repurposed-battery selling price and commercial and industrial end-user value," National Renewable Energy Lab. (NREL), Golden, CO (United States), Tech. Rep., 2012.

13 T. Montes, et al., "Procedure for assessing the suitability of battery second life applications after EV first life," *Batteries*, vol. 8, no. 9, p. 122, 2022.

14 L. C. Casals, B. A. García, and C. Canal, "Second life batteries lifespan: Rest of useful life and environmental analysis," *J. Environ. Manag.*, vol. 232, pp. 354–363, 2019.

15 S. Chai, et al., "An evaluation framework for second-life EV/PHEV battery application in power systems," *IEEE access*, vol. 9, pp. 152 430–152 441, 2021.

16 E. Michelini, P. Höschele, F. Ratz, M. Stadlbauer, W. Rom, C. Ellersdorfer, and J. Moser, "Potential and most promising second-life applications for automotive lithium-ion batteries considering technical, economic and legal aspects," *Energies*, vol. 16, no. 6, p. 2830, 2023.

17 K. Chirumalla, I. Kulkov, F. Vu, and M. Rahic, "Second life use of li-ion batteries in the heavy-duty vehicle industry: Feasibilities of remanufacturing, repurposing, and reusing approaches," *Sustain. Prod. Consum.*, vol. 42, pp. 351–366, 2023.

18 M. A. Safran, N. Badgujar, and N. Tantharatana, "Application of second life electric vehicles' lithium-ion batteries in micromobility," in *Global Conference on Sustainable Manufacturing*. Springer, pp. 847–855, 2024.

19 A. N. Patel, et al., "Lithium-ion battery second life: Pathways, challenges and outlook," *Front. Chem.*, vol. 12, p. 1358417, 2024.

20 S. Ruddell, U. K. Madawala, and D. J. Thrimawithana, "Dynamic WPT system for EV charging with integrated energy storage," *IET Power Electronics*, vol. 12, no. 10, pp. 2660–2668, 2019.

21 E. Braco, I. San Martin, P. Sanchis, A. Ursúa, and D.-I. Stroe, "State of health estimation of second-life lithium-ion batteries under real profile operation," *Appl. Energy*, vol. 326, p. 119992, 2022.

22 M. Berecibar, I. Gandiaga, I. Villarreal, N. Omar, J. Van Mierlo, and P. Van den Bossche, "Critical review of state of health estimation methods of Li-ion batteries for real applications," *Renew. Sustain. Energy Rev.*, vol. 56, pp. 572–587, 2016.

23 D.-I. Stroe, M. Swierczynski, A.-I. Stroe, R. Laerke, P. C. Kjaer, and R. Teodorescu, "Degradation behavior of lithium-ion batteries based on lifetime models and field measured frequency regulation mission profile," *IEEE Trans. Ind. Appl.*, vol. 52, no. 6, pp. 5009–5018, 2016.

24 K. A. Severson, et al., "Data-driven prediction of battery cycle life before capacity degradation," *Nat. Energy*, vol. 4, no. 5, pp. 383–391, 2019.

25 W. Pan, et al., "A health indicator extraction and optimization for capacity estimation of li-ion battery using incremental capacity curves," *J. Energy Storage*, vol. 42, p. 103072, 2021.

26 R. Klink, R. H. Van Doorn, and A. Weber, "Review–electrochemical impedance spectroscopy for lithium-ion batteries: Measurement and analysis (for automotive applications)," *Curr. Opin. Electrochem.*, p. 101768, 2025.

27 F. Larsson, S. Bertilsson, M. Furlani, I. Albinsson, and B.-E. Mellander, "Gas explosions and thermal runaways during external heating abuse of commercial lithium-ion graphite-$LiCoO_2$ cells at different levels of ageing," *J. Power Source*, vol. 373, pp. 220–231, 2018.

28 J. Huang, et al., "Predictive thermal safety of lithium-ion batteries through a unified kinetic-thermal framework," *J. Mater. Chem. A*, 2026.

29 I. E. Commission et al., "Secondary cells and batteries containing alkaline or other non-acid electrolytes - safety requirements for secondary lithium cells and batteries, for use in industrial applications" IEC 62619, 2022.

30 "Energy storage systems: Ul 1973 certification and battery components," 2023, whitepaper summarizing UL 1973 scope and examples of monitoring and safety function verification.

31 F. Turan, A. R. Boynuegri, and T. Durmaz, "Comprehensive technical and economic evaluations of using second-life batteries as energy storage in off-grid applications: A customized cost analysis," *J. Energy Storage*, vol. 120, p. 116379, 2025.

32 D. A. Notter, M. Gauch, R. Widmer, P. Wager, A. Stamp, R. Zah, and H.-J. Althaus, "Contribution of Li-ion batteries to the environmental impact of electric vehicles," 2010.

33 L. A.-W. Ellingsen, et al., "Life cycle assessment of a lithium-ion battery vehicle pack," *J. Ind. Ecol.*, vol. 18, no. 1, pp. 113–124, 2014.

34 O. Schmidt, S. Melchior, A. Hawkes, and I. Staffell, "Projecting the future levelized cost of electricity storage technologies," *Joule*, vol. 3, no. 1, pp. 81–100, 2019.

35 R. Spurrett, N. Simmons, C. Pearson, and G. Dudley, "Lithium-ion batteries based on commercial cells: Past, present and future," in *8th European Space Power Conference*, vol. 661, p. 67, 2008.

36 G. Harper, et al., "Recycling lithium-ion batteries from electric vehicles," *Nature*, vol. 575, no. 7781, pp. 75–86, 2019.

37 E. Ela, R. B. Hytowitz, and U. Helman, "Ancillary services in the united states: Technical requirements, market designs and price trends," *Electric Power Research Institute*, 2019.

38 H. Rahimi-Eichi, U. Ojha, F. Baronti, and M.-Y. Chow, "Battery management system: An overview of its application in the smart grid and electric vehicles," *IEEE Ind. Electron. Mag.*, vol. 7, no. 2, pp. 4–16, 2013.

39 S. Piller, M. Perrin, and A. Jossen, "Methods for state-of-charge determination and their applications," *J. Power Sources*, vol. 96, no. 1, pp. 113–120, 2001.

40 P. Keil, et al., "Calendar aging of lithium-ion batteries: I. impact of the graphite anode on capacity fade," *J. Electrochem. Soc.*, vol. 163, no. 9, pp. A1872–A1880, 2016.

41 N. Noura, L. Boulon, and S. Jemeï, "A review of battery state of health estimation methods: Hybrid electric vehicle challenges," *World Electr. Veh. J.*, vol. 11, no. 4, p. 66, 2020.

42 E. Martinez-Laserna, et al., "Technical viability of battery second life: A study from the ageing perspective," *IEEE Trans. Ind. Appl.*, vol. 54, no. 3, pp. 2703–2713, 2018.

43 K. Neigum and Z. Wang, "Technology, economic, and environmental analysis of second-life batteries as stationary energy storage: A review," *J. Energy Storage*, vol. 103, p. 114393, 2024.

44 B. Xu, A. Oudalov, A. Ulbig, G. Andersson, and D. S. Kirschen, "Modeling of lithium-ion battery degradation for cell life assessment," *IEEE Trans. Smart Grid*, vol. 9, no. 2, pp. 1131–1140, 2016.

45 S. F. Schuster, T. Bach, E. Fleder, J. Müller, M. Brand, G. Sextl, and A. Jossen, "Nonlinear aging characteristics of lithium-ion cells under different operational conditions," *J. Energy Storag*e, vol. 1, pp. 44–53, 2015.

7

Advanced Topics and Emerging Technologies

7.1 Introduction

Electric energy storage systems are playing an increasingly important role in the transformation of modern power systems. As electricity networks evolve from centralized structures toward more distributed, flexible, and low-carbon architectures, energy storage is no longer used only as a supporting technology. Instead, it has become a key enabler for renewable energy integration, grid reliability, and intelligent energy management. This chapter introduces advanced topics and emerging technologies that build upon the fundamental principles discussed in earlier chapters and reflect current developments in the application of energy storage in real-world power systems. A major focus of this chapter is decentralized energy storage systems (DESS), where storage is deployed close to loads, renewable generators, or within distribution networks. Unlike traditional centralized storage, decentralized systems operate under diverse conditions and often serve multiple functions at the same time. They can reduce peak demand, improve power quality, support local reliability, and increase the use of renewable energy. Understanding how these systems are designed, controlled, and coordinated is essential for modern energy storage applications.

This chapter also explores how energy storage systems are integrated with renewable energy sources such as solar photovoltaic (PV) and wind power. Solar and wind generation are clean and sustainable, but their output depends strongly on weather conditions and time of day. For example, solar power varies with sunlight and cloud cover, while wind power changes with wind speed and direction. These variations create uncertainty in power generation and make it difficult to continuously match electricity supply with demand. Energy storage systems help overcome these challenges by storing excess energy when renewable generation is high and releasing it when generation is low or demand is high. In this way, storage smooths the power output of renewable sources and improves the overall stability of the power system.

Energy storage also plays a key role in time-shifting energy, which means moving energy from periods of low demand to periods of high demand. For instance, surplus solar energy generated during midday can be stored in batteries and used during evening peak hours. In addition, modern storage technologies can respond very quickly to sudden changes in renewable generation or load. This fast response capability helps with frequency regulation, voltage support, and ramp-rate control, which are essential for maintaining power quality. As power systems move toward higher shares of renewables, the flexibility once provided by conventional fossil-fuel-based generators is reduced. In such systems, energy storage becomes a critical resource for providing both short-term power support and longer-term energy balancing.

Another major focus of this chapter is the application of energy storage in microgrids and in remote or isolated power systems. Microgrids often serve campuses, industrial sites, islands, or rural communities, where reliable access to the main grid may be limited or unavailable. In these settings, energy storage significantly improves system reliability by supplying power during generation outages or sudden load changes. Traditionally, diesel generators have been used as backup power sources in remote areas. However, frequent diesel operation leads to high fuel costs, maintenance requirements, and greenhouse gas emissions. By integrating energy storage with local renewable sources, such as solar and wind, the dependence on diesel generators can be greatly reduced. Energy storage also enables microgrids to operate in both grid-connected and islanded modes. When connected to the main grid, storage can support peak shaving, energy arbitrage, and grid services. During grid disturbances or extreme events, the microgrid can disconnect and operate independently, with storage ensuring a smooth transition and continuous power supply. This capability enhances system resilience and is increasingly important in the context of climate-related disruptions.

Overall, these applications clearly show that energy storage is not only an energy backup solution but also a powerful tool for real-time power management, system flexibility, and long-term sustainability in modern electric power systems. The integration of electric vehicles (EVs) into the power system, including vehicle-to-grid (V2G) and vehicle-to-home (V2H) concepts, is further explored. EVs introduce new challenges due to increased electricity demand, but they also offer significant opportunities because their batteries can act as mobile energy storage units. When properly managed, EVs can support grid operation, provide backup power, and participate in energy markets. This chapter discusses the control strategies and algorithms required to coordinate EV charging and discharging while meeting user and grid requirements.

7.2 Decentralized Energy Storage System

DESS have emerged as a critical component of modern power systems as electricity networks transition from centralized architectures toward distributed, low-carbon energy paradigms. Traditionally, electrical grids were designed around large, centralized generation assets with limited flexibility on the demand side. Energy storage, where present, was typically deployed at the transmission level in the form of pumped hydroelectric facilities. However, the rapid growth of distributed energy resources (DERs), particularly rooftop photovoltaics, small-scale wind generation, and electric vehicles, has fundamentally altered power flow patterns and operational requirements at the distribution level, as illustrated in Figure 7.1. In this context, decentralized energy storage provides localized balancing, buffering, and control capabilities that are increasingly essential for reliable and efficient grid operation [1].

DESS is usually deployed close to end users or embedded within distribution networks, including residential, commercial, and industrial installations. These

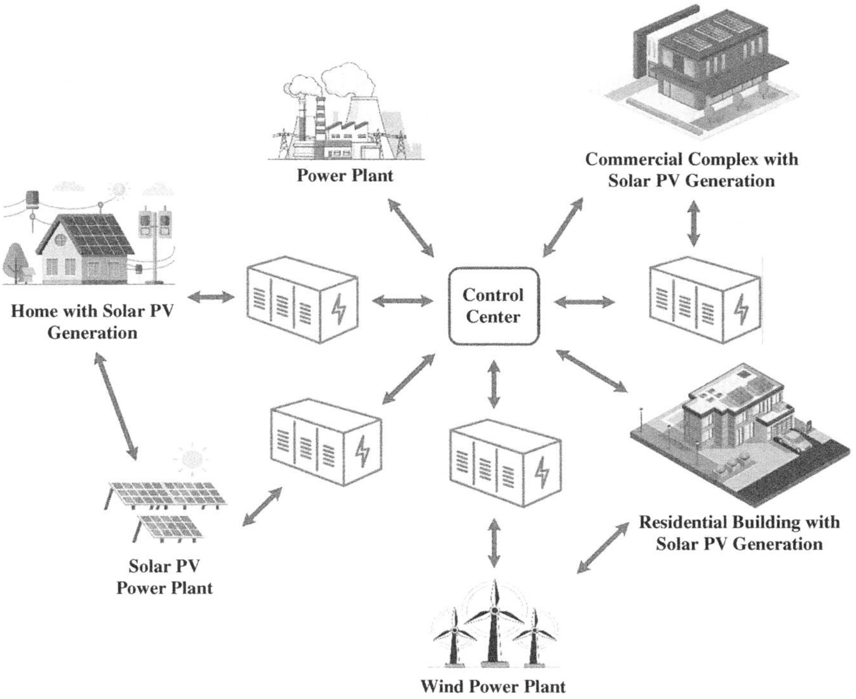

Figure 7.1 Distributed energy resources with storage system configuration.

systems can operate independently or in coordination with local generation, loads, and grid signals to manage variability, reduce peak demand, and enhance power quality. By storing energy at or near the point of consumption, decentralized storage reduces transmission losses, alleviates network congestion, and enables greater self-consumption of renewable energy. Moreover, DESS enhances system resilience by supporting islanded operation in microgrids and maintaining critical services during grid disturbances or outages.

Beyond their physical placement, decentralized energy storage systems represent a shift in how storage assets are operated and valued. Rather than serving a single, centralized function, DESS are inherently multi-functional: a single system may provide energy arbitrage, peak shaving, frequency regulation, voltage support, and backup power over different timescales. This multifunctionality introduces new technical and economic complexities, requiring advanced control strategies, communication infrastructure, and regulatory frameworks. As a result, decentralized storage is increasingly viewed not merely as an energy asset, but as an active, intelligent component of the power system.

Recent years have seen rapid advancements that significantly expand the technical feasibility and economic attractiveness of DESS. Innovations in battery chemistries, modular system design, power electronics, and digital control have enabled higher performance, improved safety, longer lifetimes, and more flexible integration with distributed resources. At the same time, the convergence of energy storage with data analytics, artificial intelligence, and networked control has enabled decentralized systems to operate collectively, delivering grid-scale services through aggregation mechanisms such as virtual power plants. These developments mark a transition from passive, standalone storage installations to coordinated, intelligent energy storage networks an evolution that is central to the future operation of low-carbon, decentralized power systems.

The critical issue that plagues decentralized energy systems that are built around green energy sources is their intermittency. This is where battery energy storage systems become a vital component in DERs.

Recent innovations in battery chemistry are expanding the performance and safety envelopes of decentralized storage. While lithium-ion technologies remain dominant due to high energy density and mature supply chains, alternative chemistries such as sodium-ion, solid-state, and flow batteries are gaining traction for decentralized applications due to lower cost, enhanced safety, and scalability for long-duration storage needs. These emerging technologies promise to mitigate raw material dependence and improve operational lifetimes compared with conventional chemistry.

It is noteworthy that combining multiple storage technologies, such as batteries, supercapacitors, or thermal systems, has shown improved adaptability to renewable generation variability and grid requirements. Hybrid systems leverage complementary strengths (e.g., high power from supercapacitors and long duration from batteries) to optimize performance, cost, and lifespan [2].

7.3 Major Components of Decentralized Energy Storage Systems

DESS are composed of multiple tightly integrated subsystems that collectively enable energy storage, power conversion, control, and interaction with the electrical grid and local loads.

Unlike centralized storage installations, DESS must operate under diverse operating conditions, respond to local constraints, and support multiple grid services. As a result, their design emphasizes modularity, flexibility, and intelligent control. The basic block diagram of DESS is shown in Figure 7.2.

The major components of a typical DESS are described below.

7.3.1 Energy Storage Medium

The energy storage medium is the core component of a DESS and determines its fundamental performance characteristics, including energy capacity, power capability, efficiency, lifetime, and safety. In contemporary decentralized applications, electrochemical batteries dominate, particularly lithium-ion technologies such as lithium iron phosphate (LFP) and nickel manganese cobalt (NMC). These chemistries offer favorable energy density, high round-trip efficiency, and mature manufacturing ecosystems, making them well-suited for residential and commercial deployments. Emerging storage media are also gaining relevance in decentralized contexts. Sodium-ion batteries offer cost and material availability advantages, while flow batteries provide scalable energy capacity and long cycle life for community-scale systems. In some applications, hybrid configurations combine batteries with supercapacitors or other storage elements to decouple energy and power functions. The choice of storage medium is closely linked to the intended application, duty cycle, and economic constraints of the DESS.

7.3.2 Power Conversion System (PCS)

The power conversion system interfaces the energy storage medium with the electrical grid and local loads. It typically consists of bidirectional DC-to-DC converters, DC-to-AC inverters, and associated protection and filtering elements. The PCS governs charge and discharge processes, regulates voltage and frequency, and ensures compliance with grid interconnection standards.

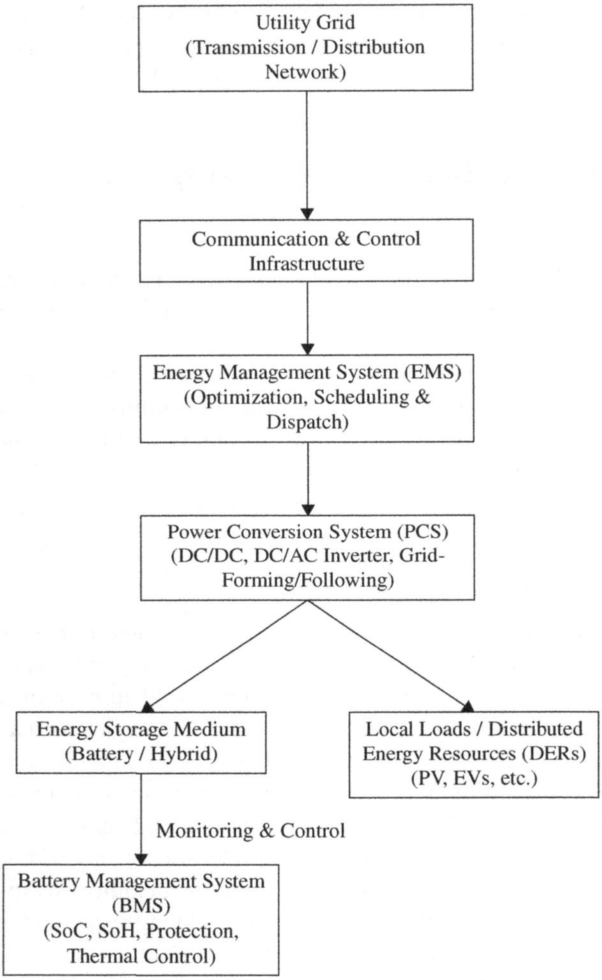

Figure 7.2 Block diagram of a decentralized energy storage system (DESS).

In decentralized systems, the PCS must accommodate a wide range of operating modes, including grid-connected operation, islanded or microgrid operation, and seamless transitions between these states. Advanced inverter functionalities, often referred to as grid-forming or grid-following capabilities, allow DESS to provide services such as voltage regulation, frequency support, and fault ride-through. The increasing intelligence embedded in power electronic interfaces has elevated the PCS from a passive conversion stage to an active grid-interactive component.

7.3.3 Battery Management System (BMS)

The battery management system is responsible for ensuring the safe and reliable operation of the energy storage medium. Core BMS functions include monitoring cell voltages, currents, and temperatures; estimating state of charge (SoC) and state of health (SoH); balancing cells; and enforcing operational limits to prevent degradation or failure.

In decentralized deployments, the BMS plays a critical role in extending system lifetime and enabling second-life battery applications. Advanced BMS architectures increasingly incorporate data-driven models and adaptive algorithms to account for aging, environmental variability, and heterogeneous cell behavior. The BMS also serves as a key interface between the physical battery system and higher-level energy management and control layers.

7.3.4 Energy Management System (EMS)

The energy management system coordinates the operation of the DESS in response to local objectives and external signals. The EMS determines when and how much energy to store or release based on factors such as load demand, renewable generation, electricity prices, and grid service requirements. In decentralized settings, the EMS must often balance competing objectives, including cost minimization, self-consumption maximization, emissions reduction, and resilience enhancement. Modern EMS implementations leverage forecasting, optimization, and real-time control techniques. When multiple decentralized storage units are aggregated, the EMS may operate hierarchically or in a distributed manner, enabling coordinated behavior across a neighborhood, campus, or virtual power plant. This layer is central to unlocking the multifunctional value of DESS.

7.3.5 Communication and Control Infrastructure

Effective operation of DESS relies on robust communication and control infrastructure that links sensors, controllers, inverters, and external entities such as utilities or market operators. Communication protocols enable data exchange for monitoring, dispatch, and coordination, while control algorithms translate high-level objectives into real-time operational commands.

In decentralized environments, this infrastructure must support interoperability across devices from different manufacturers and maintain cybersecurity and data integrity. Increasingly, decentralized and peer-to-peer control approaches are being explored to reduce reliance on centralized coordination and enhance system scalability and resilience.

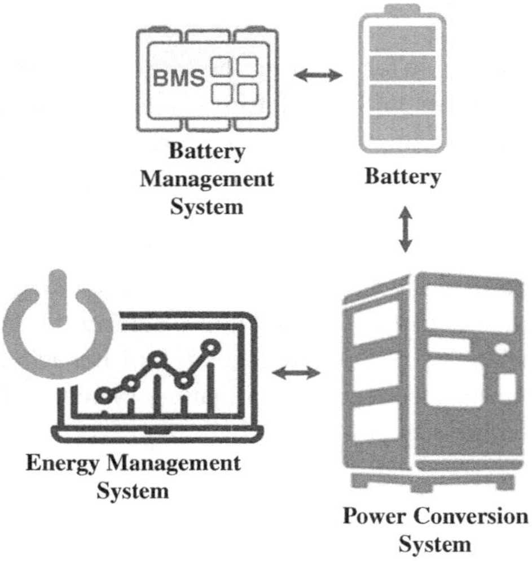

Figure 7.3 Basic configuration of DESS.

7.3.6 Protection, Safety, and Auxiliary Systems

DESS installations incorporate a range of auxiliary components to ensure safe and compliant operation. These include electrical protection devices, thermal management systems, fire detection and suppression equipment, and mechanical enclosures. Safety considerations are particularly important in decentralized installations located in residential or commercial buildings, where space constraints and human proximity impose strict design requirements. Thermal management systems are critical for maintaining performance and preventing accelerated degradation or thermal runaway in battery-based storage systems. Together, these auxiliary systems ensure that DESS can operate reliably across diverse environmental and load conditions. The DESS configuration is depicted in Figure 7.3.

7.4 Detailed Insights into Core Applications

7.4.1 Microgrids: Resilient Local Ecosystems

Microgrids are self-contained energy networks that can operate independently from the main utility grid, a mode known as "islanding."

- **Operational Independence:** By integrating renewable sources (solar and wind) with battery energy storage systems (BESS), microgrids can maintain power during mass grid outages or extreme weather.

  Diverse Users: Beyond remote communities, they are increasingly used by military bases, hospitals, and industrial logistics terminals—such as Puerto

Rico's Isla Grande terminal-to ensure seamless operations regardless of external grid conditions.

- **Smart Control:** Modern microgrids use AI-powered grid controllers to forecast demand and balance loads in real time, reducing energy waste and providing financial benefits to onsite operators.

7.4.2 Virtual Power Plants (VPPs): Digital Power Stations

A VPP is a digital network that aggregates thousands of decentralized assets, such as home batteries, rooftop solar, and smart thermostats, to function as a single large-scale power plant.

- **Grid Services:** VPPs participate in energy markets by providing "ancillary services," such as frequency regulation and voltage support, which help stabilize the grid faster than traditional fossil-fuel plants.
- **Revenue Generation:** Participants can earn income through storing energy when it is cheap (or abundant from solar) and selling it back during peak demand periods when prices spike.
- **Market Impact:** In the United States, VPP capacity is estimated between 30 and 60 GW, with projects like the South Australia VPP, which is scaling to connect to 50 000 homes.

7.4.3 Prosumer Integration and Peer-to-peer (P2P) Trading

Decentralization empowers prosumers, who are individuals that both produce and consume energy and can interact directly with other users.

- **Local Markets:** P2P trading platforms create marketplaces where neighbors can trade renewable energy without traditional intermediaries. This can reduce electricity bills for consumers while providing better returns for prosumers than standard feed-in tariffs.
- **Community Economy:** Research suggests that communities enabled with P2P trading achieve higher energy self-sufficiency than those focusing only on individual self-consumption.
- **Dynamic Pricing:** The rollout of smart meters is expected to facilitate dynamic pricing, allowing households to sell excess energy automatically whenever retail prices exceed set thresholds.

7.4.4 Vehicle-to-grid (V2G) Technology

V2G transforms EVs from passive loads into mobile storage units that can feed electricity back into the grid.

- **Fleet Anchoring:** Public transportation with their predictable schedules and large batteries allow them to provide significant grid support when they are not in use.
- **Fast-response Balancing:** While stationary batteries dominate total capacity, EV batteries are critical for rapid-response balancing and localized backup power during outages.

Decentralized storage architectures are increasingly modular and grid-aware. Distributed control algorithms and peer-to-peer coordination strategies allow clusters of storage units (e.g., community batteries) to optimize charging and discharging without centralized command, improving responsiveness and maintaining system stability in low-voltage networks [3]. Continued research on distributed optimization and consensus-based control strategies aims to address these issues. Progress in materials science, system integration, and hybrid technologies will likely shape the next generation of decentralized energy storage solutions.

7.5 Integration of Energy Storage with Renewable Energy Systems

Energy storage systems (ESS) play a crucial role in managing the variability and uncertainty associated with renewable energy sources. Solar, wind, and other renewable resources do not produce power continuously, as their output depends on environmental conditions such as sunlight and wind speed. This intermittency can create mismatches between electricity generation and demand. ESS helps to address this challenge by storing excess energy when renewable generation is high and supplying it during periods of low generation or sudden interruptions. In this way, energy storage smooths power fluctuations and supports the stable operation of power systems with high shares of renewable energy [4–6].

Battery-based ESS is widely used because it can respond very quickly to changes in load or generation. Batteries are capable of rapidly absorbing power when generation exceeds demand and injecting power when demand increases. This fast response makes them suitable for important grid-support functions such as voltage regulation, frequency control, and peak shaving. Peak shaving reduces the maximum power drawn from the grid during high-demand periods, which lowers stress on network infrastructure and reduces operating costs. By integrating ESS into renewable power plants, system reliability is improved, and higher levels of renewable penetration can be achieved without compromising grid stability. The practical integration of ESS with renewable energy systems requires suitable power-conversion hardware and coordinated control strategies. Key components include inverters, DC-to-DC converters, maximum power point tracking (MPPT) controllers, and protection devices. These components ensure efficient energy transfer between renewable generators, storage units, and the grid. Control strategies are designed to coordinate the operation of the ESS with renewable generation and grid requirements, allowing smooth power flow under both steady-state and transient conditions. Recent research also highlights the advantages of hybrid energy storage configurations, where multiple storage technologies are combined to exploit their complementary characteristics. For example, pairing lithium-ion batteries with supercapacitors allows the system to achieve both high energy density and fast power response. Such hybrid systems have been shown to improve efficiency, reduce battery stress, and enhance dynamic performance in renewable applications [7, 8].

Modern renewable energy systems increasingly adopt hybrid AC/DC microgrid architectures to reduce conversion losses and improve overall efficiency, as illustrated in Figure 7.4. In these architectures, renewable sources such as solar PV, wind turbines with DC outputs, and battery banks are connected to a common DC bus whenever possible. This arrangement minimizes the number of AC-to-DC and DC-to-AC conversion stages, which reduces energy losses and simplifies system control. The DC bus is then connected to the main AC grid through inverters and bidirectional converters, allowing seamless interaction with conventional AC networks. At the same time, local DC loads, such as data centers, LED lighting, and electric vehicle charging stations, can be supplied directly from the DC bus [9–11].

Battery management systems are essential for the safe and efficient operation of battery-based ESS. A BMS continuously monitors key parameters such as state of charge, depth of discharge, voltage, current, and temperature. By enforcing operational limits, the BMS protects the battery from overcharging, deep discharging, and thermal stress, thereby extending battery life and improving system reliability. Advanced energy management controllers increasingly use artificial

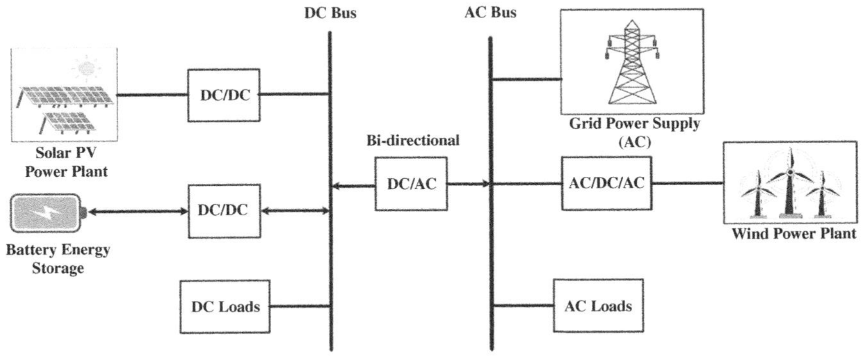

Figure 7.4 Hybrid AC/DC microgrid architecture.

intelligence and predictive algorithms to optimize charging and discharging schedules. These algorithms take into account electricity prices, weather forecasts, and load patterns to support cost-effective operation and grid services [12].

Practical examples from recent literature show the economic benefits of integrating ESS with renewable generation [13]. For instance, when a battery system is coupled with a wind farm, it can reduce imbalance penalty costs by approximately 15–40% by compensating for forecast errors and smoothing power output. In addition, strategic dispatch of stored energy for peak shaving and time-shifting can increase overall revenue by around 8–10%. These results clearly demonstrate that energy storage is not only a technical solution for variability but also a valuable asset for improving the economic performance of renewable energy systems. Table 7.1 summarizes key storage options for renewable integration.

Table 7.1 Key storage options for renewable integration.

Storage technology	Key characteristics	Typical applications
Lithium-ion batteries	High energy and power density; fast response; modular. Subject to cost, degradation, and resource issues.	Grid-scale smoothing; solar/wind farms, EV charging support.
Flow batteries	Scalable energy capacity; long cycle life; flexible output. Lower power density and higher complexity.	Utility-level renewable buffering; long-duration storage.
Pumped hydro storage	Very large capacity; proven reliability; low cost per kWh. Geographically constrained.	Bulk grid storage; peak load shifting; seasonal storage.
Compressed air (CAES)	Large-scale storage; long discharge. Requires suitable geology; efficiency losses in compression.	Grid resilience; renewable firming.
Supercapacitors	Extremely high power density; instant response. Very limited energy capacity.	Power quality, smoothing rapid fluctuations in renewables.
Thermal storage	High capacity for heat or heat-to-power; e.g., molten salt. Suitable for CSP or building HVAC; material costs can be high.	Concentrated solar (CSP), waste heat recovery, district heating.
Hydrogen/synthetic fuels	Very high energy per mass; transports energy over long time/distances. Round-trip efficiency low (due to conversion losses).	Seasonal storage; power-to-gas, renewable dispatchable fuels.
Flywheels	High power and long cycle life; fast dynamic response. Energy leaks over hours (self-discharge via friction).	Frequency regulation; UPS and uninterruptible power backups.

Different storage technologies serve different roles in renewable integration. Li-ion batteries and vanadium flow batteries offer high energy density and scalability for grid-scale buffering, even though at a higher cost and with resource constraints. Supercapacitors and flywheels provide a very fast response for power conditioning but hold little total energy. Pumped hydro and compressed-air storage deliver massive capacity suitable for seasonal or bulk storage, though they require specific sites. Thermal storage excels at coupling with heat processes, and hydrogen or synthetic fuels enable long-duration storage across seasons at efficiency costs. Recent reviews stress that hybridizing multiple storage types often yields the best performance, as a mix of batteries, capacitors, and mechanical storage can better cover the spectrum of power and energy needs.

Including energy storage also reshapes system operations. For instance, in grid-tied systems, ESS can engage in energy arbitrage, such as storing energy when prices are low and selling when high and provide ancillary services such as frequency/voltage regulation. In distribution networks, optimally scheduling battery charging/discharging can improve voltage profiles and reduce losses. Hybrid renewable-storage plants like solar, wind, and battery are being explored with AI-driven controls to maximize efficiency and minimize costs. Overall, recent literature emphasizes that effective integration relies on carefully co-designing the generation and storage control. Advanced inverters, multilevel converters, and microgrid controllers are needed to seamlessly tie together renewables, storage, and the grid.

7.6 Application of Energy Storage in Microgrids and Remote Areas

A microgrid is a local power network that can operate connected to the main grid or in island mode. It typically contains distributed renewable sources such as solar PV, wind, small hydro, etc., conventional generators or storage, and loads under a common controller (Figure 7.5). ESS are fundamental in microgrids, especially those in remote or off-grid areas. They buffer the variability of renewables, allowing the microgrid to supply stable power to local users. As noted in recent reviews, "energy storage systems are essential elements that provide reliability and stability in microgrids with high penetrations of renewable energy sources". In a remote village microgrid, batteries or other storage can take up the slack when solar drops at evening, supplying the load until sunrise. This capability dramatically increases energy access and quality for communities otherwise reliant on diesel generators or no power.

The technical benefits of storage in a microgrid are many. ESS allow greater use of local renewables such as solar and wind by absorbing excess generation and

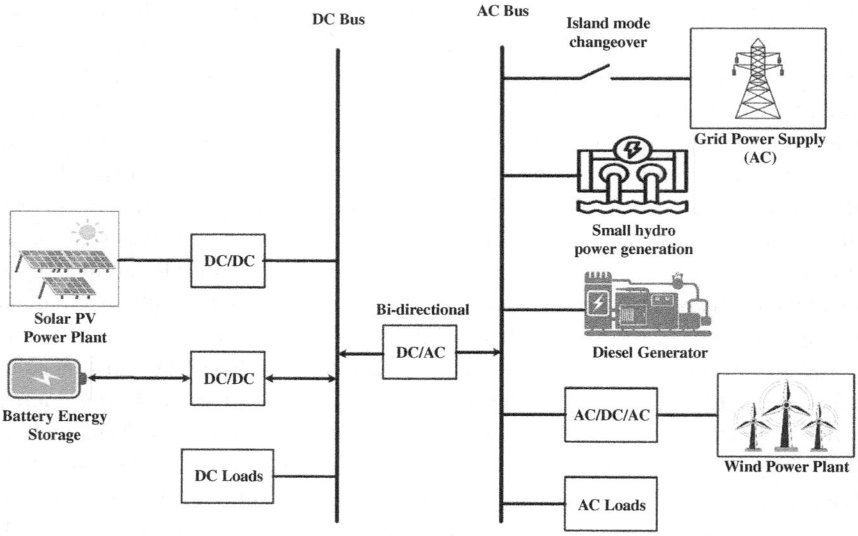

Figure 7.5 Microgrid architecture with grid-connected and islanded modes of operation.

reducing curtailment. They also handle load changes and provide fast response to disturbances, stabilizing frequency and voltage when isolated from the main grid. For example, in islanded microgrids, batteries can quickly inject or absorb power to keep the system balanced. Moreover, intelligently managed ESS can defer expensive diesel use, batteries charge from solar by day and discharge at night or peak times, lowering fuel costs and emissions. Studies showed that integrating renewables with diesel and BESS in an isolated microgrid improves reliability and cuts operational costs, underlining the importance of well-coordinated dispatch.

In many microgrids, both AC and DC loads exist such as lights, appliances vs. DC motors or USB-powered devices, so hybrid AC/DC designs can improve efficiency by matching sources to loads. The battery energy storage system (BESS) often occupies the DC side, charging via an inverter or DC-to-DC converter. A microgrid controller or energy management system coordinates the BESS with generation forecasts to maintain balance.

In remote or rural electrification projects, examples abound of ESS enabling clean microgrids. Studies emphasize that off-grid storage mitigates the intermittency and volatility of renewable energy generation and provides support during peak demand, greatly improving stability. In fact, a recent analysis found that a well-chosen storage configuration can significantly enhance the utilization rate of renewable energy, reduce system operating costs, and strengthen grid resilience under extreme conditions. Many projects use lead-acid or lithium batteries paired with PV. Hybrid storage combinations like batteries plus supercapacitors or compressed air have been proposed for microgrids to exploit the quick power of capacitors and the longer-duration storage of batteries.

The practical operation of microgrids with storage also requires smart control. Controllers must manage state-of-charge, schedule battery cycling to extend life, and handle transitions between grid-connected and islanded modes. Microgrids are often classified as AC, DC, or hybrid (combining both) systems, and storage integration strategies depend on the architecture. For AC microgrids, batteries connect through inverters, for DC microgrids, batteries tie directly into the DC bus with appropriate converters. Under high renewable penetration, advanced controls, even AI-based are used to predict generation and load, optimizing storage dispatch [11–13].

Key advantages of ESS in microgrids include:

- **Enhanced Reliability:** Storage provides backup during outages or supply gaps, ensuring a continuous power supply.
- **Renewable Smoothing:** Batteries absorb short-term fluctuations from wind/PV, allowing smoother power delivery.
- **Peak Shaving:** By discharging during peak load, ESS reduce stress on generators and lines.
- **Reduced Fuel Use:** In diesel-based microgrids, solar-plus-storage can dramatically cut generator run-time.
- **Economic Operation:** Properly dispatched storage can minimize operating costs (e.g., diesel/fuel) and improve asset utilization.

In summary, integrating energy storage with renewables whether at utility scale or in microgrids transforms how power systems operate. Storage enables higher renewable penetration, grid stability, and flexible operation. In remote and rural microgrids, storage is the essential component that turns intermittent generators into reliable power sources, greatly expanding energy access while reducing costs and emissions. As current literature shows, both the hardware (batteries, converters, hybrid configurations) and the control algorithms (optimal dispatch, hybrid management) continue to advance rapidly, promising even more resilient and efficient renewable energy systems in the near future.

7.7 Integration of EVs into Grid

EVs are transforming how electricity systems operate by introducing a new, large, and flexible load into the power grid. Unlike traditional loads, EVs are equipped with batteries that can not only consume energy but also, under the right conditions, supply energy back to the grid. This two-way interaction between EVs and the

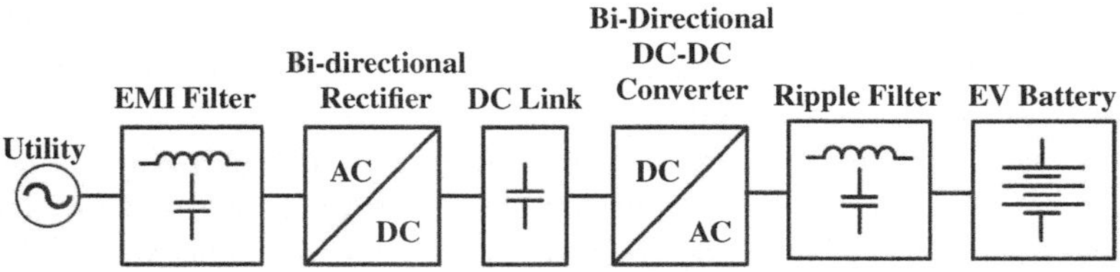

Figure 7.6 Grid-integrated EV charging system.

electricity network, referred to as vehicle-to-grid (V2G), represents a major opportunity for enhancing grid flexibility, reliability, and renewable energy utilization.

At its simplest level, grid integration of EVs involves connecting vehicles to the power system through advanced charging infrastructure that supports bidirectional power flow, as depicted in Figure 7.6. In a V2G setup, EV batteries act as distributed energy storage units, absorbing electricity during periods of low demand or high generation and delivering energy back to the grid during peak demand or when supply is constrained. This capability means that large fleets of EVs can collectively provide ancillary services such as frequency regulation, voltage support, and peak shaving, traditionally supplied by dedicated storage or fast-response generators.

Integrating EVs with the grid also plays a critical role in accommodating higher shares of variable renewable energy sources like solar and wind. Renewables can generate excess energy during certain times of day, which can be stored in EV batteries and later discharged when generation drops or demand spikes. By doing so, EV integration helps smooth supply fluctuations and reduces the need for fossil-fuel backup generation. In addition, managed or smart charging, where EV charging is coordinated with grid conditions, can shift EV load to off-peak periods, lowering overall system costs and reducing stress on distribution networks.

Despite these benefits, integrating EVs at scale presents technical, economic, and regulatory challenges. One major technical requirement is the widespread deployment of bidirectional chargers and communication systems that enable real-time coordination between vehicles, utilities, and grid operators. Without standardized protocols and interoperable hardware, seamless integration remains difficult. Battery degradation is another concern; frequent cycling associated with V2G operations can accelerate wear and reduce battery life if not properly managed, though impacts vary with chemistry and use patterns. Moreover, cybersecurity risk increases as EV charging infrastructure becomes more connected, necessitating robust protections to safeguard grid and user systems.

Several countries and utilities are already exploring grid-integrated EV pilots to test the operational and economic viability of V2G services. Research continues into optimization algorithms and control strategies that balance user driving needs with grid flexibility services, often leveraging predictive forecasts and real-time price signals to coordinate charging and discharging efficiently.

One of the primary concerns in EV grid integration is the impact of uncoordinated charging. When many EVs begin charging simultaneously (often during evening hours when residential demand is already high) resulting in a load increase, exacerbating the peak demand, overloading distribution transformers, and causing voltage deviations in low-voltage networks

To address these challenges, smart or controlled charging is identified as a key enabling technology. Smart charging adjusts the timing and rate of EV charging based on grid conditions, electricity prices, or renewable energy availability. By shifting charging to off-peak periods or periods of high renewable generation, smart charging can reduce peak loads, improve grid asset utilization, and lower overall system costs. Importantly, many of these benefits can be achieved without bidirectional power flow, making smart charging a practical and scalable first step toward large-scale EV integration.

The integration of EVs also shows strong synergy with renewable energy sources. Variable renewable generation, such as solar and wind power, introduces fluctuations that must be balanced in real time. EV batteries can absorb excess renewable energy during periods of high generation and release it during periods of low output or high demand. Coordinated charging and V2G operation therefore help reduce renewable energy curtailment and smooth net load profiles, supporting higher levels of renewable penetration in the power system.

Despite these advantages, there are several remaining challenges. Battery degradation is a major concern, as increased cycling associated with V2G operation can accelerate battery aging if not properly managed. Effective control algorithms must therefore account for battery health constraints and limit excessive depth of discharge. In addition, widespread EV grid integration requires reliable communication infrastructure, interoperable standards, and robust cybersecurity measures to ensure safe and secure operation as EVs become active participants in grid control [14].

As EV adoption continues to grow, their role as both transportation assets and distributed energy storage resources is expected to become an integral part of future smart grids. When aggregated, EVs function as a DESS, providing essential flexibility to manage the intermittency of renewable energy sources (RES) like wind and solar, as discussed above.

In addition to system-level impacts, EV integration introduces important challenges at the distribution network level. The ability of low-voltage feeders to accommodate EV charging (hosting capacity) is limited by voltage constraints, thermal limits, and phase imbalance. These issues are particularly pronounced with high-power fast chargers and clustered residential adoption. Addressing such constraints requires hierarchical control strategies that operate across multiple timescales and account for both technical limits and user mobility behavior. Market-based coordination mechanisms, such as dynamic pricing and flexibility markets, are increasingly viewed as scalable solutions that reduce communication complexity while preserving user autonomy [15].

7.7.1 Technical Challenges of Large-scale EV Integration

Uncoordinated or "random" charging poses substantial risks to the distribution grid, particularly in low-voltage sections.

- **Grid Stability and Overloading:** Simultaneous high-power demand from many EVs can exceed the rated capacity of distribution transformers, potentially shortening their lifespan or leading to catastrophic failures.
- **Voltage and Frequency Deviations:** High concentrations of EVs charging, especially using fast-charging methods, can cause significant voltage drops or fluctuations. In severe cases, these deviations can lead to voltage collapse and widespread blackouts.
- **Power Quality Degradation:** EV chargers act as non-linear loads, introducing harmonic distortions (varying from 3% to 28% during charging) into the system. This can lead to equipment malfunctions, increased thermal losses, and interference with power-line communications.
- **Phase Imbalance:** The random connection of single-phase chargers to three-phase distribution systems can cause severe phase unbalance, leading to negative sequence currents that damage induction motors and increase resistive losses in lines.
- **Grid Congestion:** Research suggests that transmission capacity is a primary bottleneck; without coordinated upgrades, the climate benefits of electrification may be stymied by the grid's inability to deliver renewable power to high-demand EV clusters [16].

Strategic Integration and Mitigation Frameworks

To address these challenges, recent research emphasizes the transition from uncoordinated charging to advanced, multi-objective optimization frameworks.

- **Coordinated Charging Strategies:** Implementing hierarchical scheduling can significantly reduce grid stress. For instance, studies have shown that coordinated strategies can reduce grid load variance by up to 64% and user costs by 21% compared to uncoordinated methods.
- **Optimization Algorithms:** Advanced algorithms such as particle swarm optimization (PSO) and genetic algorithms (GA) are used to determine the optimal placement and capacity of charging infrastructure. These frameworks can reduce power losses by approximately 49.56% and improve voltage stability indices by over 43%.

- **Renewable Energy Integration:** Combining EV charging with DERs, like solar photovoltaics, adds flexibility to the system. EVs can function as storage for intermittent renewable sources, minimizing their dependency on the main grid.

The integration of EVs into the smart grid is categorized by the direction and intelligence of power flow.

- **Grid-to-vehicle (G2V):** The traditional unidirectional flow where EVs are treated as passive loads. Uncoordinated G2V charging (often called "dumb charging") can exacerbate peak demand and lead to transformer overloading and voltage instability [17].
- **Managed/Smart Charging (V1G):** A unidirectional but controlled process where charging rates are adjusted based on grid signals, such as time-of-use (ToU) tariffs or frequency deviations, to shift load away from peak periods [18].
- **Vehicle-to-everything (V2X):** A bidirectional framework where EVs can discharge energy back to the grid (V2G), to a home (V2H), or to a building (V2B). This allows the vehicle battery to act as a buffer, providing ancillary services such as frequency regulation and spinning reserves [19].

7.7.2 Vehicle-to-grid Application

Vehicle-to-grid (V2G) systems extend the role of EVs beyond transportation by enabling bidirectional power exchange between EV batteries and the electric grid. In a V2G framework, EVs can both draw energy from the grid for charging and supply stored energy back to the grid when needed, as shown in Figure 7.7. This capability allows EVs to function as distributed energy storage resources, supporting grid operation while creating additional value streams for vehicle owners and system operators [20].

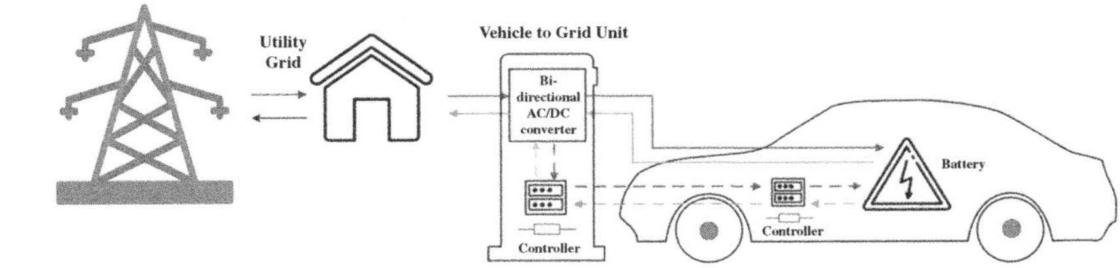

Figure 7.7 Vehicle-to-grid (V2G) system.

- **Operating Principle of V2G:** The basic operation of a V2G system relies on bidirectional chargers, power electronic interfaces, and communication infrastructure that coordinates charging and discharging actions. When connected to the grid, an EV can respond to external control signals or price incentives to either absorb power or inject power into the grid. Individually, a single EV contributes limited power, but when aggregated across large fleets, EVs can collectively provide services comparable to stationary energy storage systems. This aggregation is a central concept in V2G deployment and enables participation in grid-level services.
- **Grid Services Enabled by V2G:** V2G systems can provide a range of grid services across different timescales. At short timescales, EVs can offer fast-response services such as frequency regulation and voltage support due to the rapid controllability of power electronic converters. At longer timescales, V2G can support peak shaving, load balancing, and energy arbitrage by shifting energy from low-demand periods to peak-demand periods.
- **Aggregation and Control Strategies:** Effective V2G operation requires coordinated control of large numbers of EVs through aggregators or centralized management platforms. Control strategies must account for grid requirements, battery constraints, and user mobility needs. The hierarchical and optimization-based control approaches, where high-level controllers determine aggregate power targets while lower-level controllers manage individual EV charging and discharging actions, are critical. These strategies enable scalable deployment while ensuring that vehicle availability and user preferences are respected.
- **Battery Degradation and User Considerations:** Battery degradation is a key concern in V2G systems, as additional charge–discharge cycles can accelerate battery aging if not properly managed. The V2G control algorithms must limit depth of discharge, cycling frequency, and operating temperature to mitigate degradation effects. Furthermore, economic incentives must adequately compensate vehicle owners for battery wear and participation in grid services. User acceptance is strongly influenced by guarantees on minimum state of charge and mobility availability, making these constraints essential components of practical V2G implementations.
- **Infrastructure, Standards, and Challenges:** The widespread adoption of V2G depends on the availability of bidirectional charging infrastructure, standardized communication protocols, and supportive regulatory frameworks. Interoperability between vehicles, chargers, aggregators, and grid operators remains a major challenge. The lack of harmonized standards and market mechanisms has limited large-scale V2G deployment, despite successful pilot projects. Cybersecurity and data privacy are also critical concerns as EVs become active participants in grid control.

7.8 Vehicle-to-home Application

Vehicle-to-home (V2H) systems enable bidirectional power flow between an EV and a residential electrical system. In a V2H configuration, the battery of an EV supplies electricity directly to a home, allowing the vehicle to function as a mobile energy storage unit. Unlike V2G systems, which focus on grid-level services, V2H is primarily designed to support household energy management, backup power, and increased utilization of local renewable energy sources (Figure 7.8).

Operating Principle of V2H: A V2H system relies on a bidirectional charger and a home energy management system (HEMS) to control power exchange between the EV battery and the household electrical network. When connected, the EV can charge from the grid or from on-site renewable generation, such as rooftop solar PV systems. During periods of high household demand, grid outages, or high electricity prices, the EV battery can discharge energy to supply household loads. This operation effectively extends the functionality of stationary residential battery storage by leveraging the larger capacity of EV batteries.

Applications and Benefits: One of the most important applications of V2H is **a** backup power supply. During grid outages, an EV equipped with V2H capability can maintain critical household loads such as lighting, refrigeration, and communication systems for extended periods. Compared to conventional backup generators, V2H systems operate silently, produce no local emissions, and require no additional fuel storage.

V2H also supports energy cost reduction and self-consumption optimization. By charging the EV during off-peak periods or when excess solar generation is available, and discharging during peak demand periods, households can reduce electricity bills and reliance on the grid. In homes with renewable generation, V2H increases the local use of renewable energy and reduces curtailment, similar to stationary battery systems but with greater energy capacity.

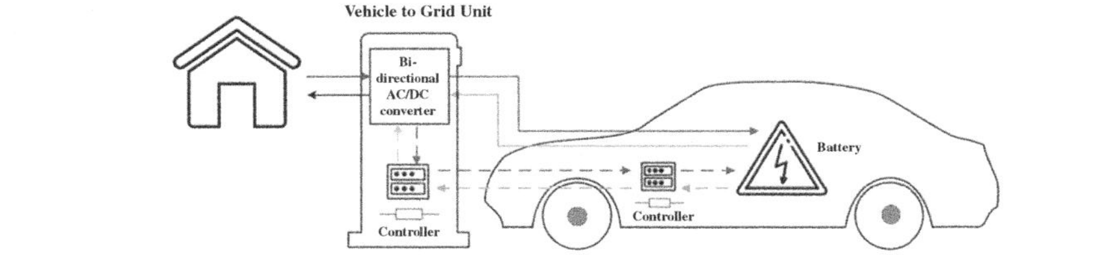

Figure 7.8 Vehicle-to-home (V2H) system.

7.8.1 Control and Energy Management

Effective V2H operation requires coordinated control strategies that balance household energy needs with vehicle availability. Home energy management systems typically prioritize mobility requirements by maintaining a minimum state of charge in the EV battery. Control algorithms may incorporate electricity prices, load forecasts, and renewable generation forecasts to determine optimal charging and discharging schedules. These strategies are generally less complex than V2G control, as they involve a single household rather than large-scale aggregation.

7.8.2 Battery Degradation and User Considerations

As with V2G, battery degradation is an important consideration in V2H systems. Additional cycling of the EV battery can accelerate aging if not properly managed. However, V2H applications often involve lower power levels and fewer cycles compared to grid services, which can reduce degradation impacts. User acceptance of V2H is strongly influenced by guarantees on vehicle readiness, simplicity of operation, and clear economic benefits. Ensuring that household energy use does not compromise driving needs is essential for widespread adoption.

Infrastructure and Standards

The deployment of V2H systems depends on the availability of bidirectional chargers, compatible vehicle architecture, and appropriate electrical standards. Several manufacturers have begun offering V2H-capable vehicles and chargers, particularly in regions with high residential solar adoption. Safety, protection coordination, and compliance with local electrical codes are critical aspects of V2H implementation, especially during islanded operation when the home is electrically separated from the grid.

7.8.3 Related Algorithms

Algorithms associated with energy storage are exhaustive. It can be classified into different verticals with multiple combinations under the umbrellas of application to technology used. To get the grasp of the picture, this section captures the applications that are key to energy storage, and only briefly touches upon the technology that powers them, owing to the scope of the chapter.

Energy Management and Scheduling Algorithms: These algorithms determine when and how much energy must be consumed/supplied by the energy storage device. The typical objectives include cost minimization, peak shaving, self-consumption maximization, and lifetime extension. It can be implemented using simple rule-based scheduling or time-of-use/price-signal-based scheduling. Some applications see the use of deterministic or even stochastic optimization, sometimes touching upon model predictive control.

When it comes to EVs, smart charging algorithms take the forefront that take into consideration load shifting, valley filling, and network-aware charging. Aggregation and fleet management algorithms are also crucial in this context to enable EVs to act collectively. This entails a centralized aggregator-based optimization, control architecture, and distributed fleet coordination.

In the context of V2G and V2H, algorithms enable EVs to provide grid services/home support. These include bidirectional power dispatch optimization, coordinated energy scheduling including DERs, frequency regulation control for grid-forming inverters, home energy management optimization and blackout support, and constrained operation algorithms. When multiple units are involved, there is an opportunity to leverage centralized or decentralized control based on actual application requirements.

Grid Integration Algorithms: These are critical components that enable grid tied operation of energy storage devices. Namely, droop control, Volt-VAR control, optimal power flow, network constrained dispatch algorithms. Economic factors also play a key role here, with predominant factors like dynamic pricing, real-time tariffs, auction-based and flexibility markets and in the inception stages, incentive-based participation models. For microgrids, islanding and resynchronization algorithms play an important role in seamless transition of control employing load shedding and restoration algorithms, when required.

Battery Health and Degradation Aware Algorithms: Unlike traditional power system assets, electrochemical batteries experience irreversible degradation with use, which directly affects their capacity, efficiency, safety, and lifetime cost. Battery health and degradation-aware algorithms are therefore essential for ensuring the long-term viability of energy storage–based solutions. Accurate estimation of battery state variables is a prerequisite for health-aware operation. In addition to SoC, advanced algorithms estimate SoH, internal resistance, and remaining useful life. These estimates are obtained using model-based observers, filtering techniques, or data-driven approaches. Traditional energy management algorithms optimize storage operation based on short-term objectives such as cost minimization or peak reduction, often neglecting long-term battery health. Degradation-aware dispatch algorithms explicitly incorporate battery aging into the optimization objective or constraints. This is commonly achieved by assigning an economic cost to battery degradation or by limiting harmful operating conditions such as deep cycling or high C-rate operation. In decentralized energy storage systems and EV-based storage, these approaches allow operators to balance immediate operational benefits against long-term asset preservation.

Supplementary Algorithms: Other than the algorithms mentioned above, it is important to note some other areas identified with active research going on associated with energy storage, which are algorithms for renewable energy resources forecasting, short-term and long-term load forecasting, machine learning algorithms for control and dispatch involving hybrid physics-informed learning models.

Another key component that needs to be taken into account is cybersecurity and fault detection. This involves anomaly and intrusion detection systems, building up to fault diagnosis and self-healing control.

Algorithm class	Primary application (DESS/EV/V2X)	Key functions
Rule-based and heuristic control	DESS, EV charging, V2H	Simple scheduling, peak shaving, backup operation
Deterministic optimization (LP/MILP)	DESS dispatch, EV charging, V2G scheduling	Cost minimization, constraint handling, unit commitment
Stochastic optimization	DESS with renewables, EV fleets	Uncertainty in load, renewables, mobility
Model predictive control (MPC)	DESS EMS, EV smart charging, V2H	Multi-timescale control, forecast-based dispatch
Distributed optimization (ADMM)	DESS coordination, EV aggregation	Scalability, privacy-preserving coordination
Consensus-based control	DESS networks, EV fleets	Power sharing, decentralized coordination
Droop and grid-forming control	Storage in microgrids, V2G inverters	Voltage/frequency regulation, islanded operation
Optimal power flow (OPF/D-OPF)	DER and storage grid integration, EV-aware networks	Network-constrained dispatch, voltage control
Smart charging algorithms	EV grid integration	Valley filling, congestion avoidance
V2G dispatch and ancillary services	EVs as storage (V2G)	Frequency regulation, reserves
Degradation-aware control	DESS, V2G, V2H	Battery lifetime optimization
Mobility-aware EV algorithms	EVs, V2G	User constraints, availability modeling
Market-based and price-responsive control	DESS, EVs, V2X	Economic coordination, flexibility markets
Reinforcement learning (RL)	Emerging DESS and V2X control	Adaptive, model-free optimization
Forecasting (ML and statistical)	DESS EMS, EV charging	Load, renewable, mobility prediction

References

1 L. Caliman, “What does the future hold for energy storage and decentralised networks?” *Polytechnique Insights*, 2025.

2 A. M. Adeyinka, O. C. Esan, A. O. Ijaola, and P. K. Farayibi, “Advancements in hybrid energy storage systems for enhancing renewable energy-to-grid integration,” *Sustain. Energy Res.*, vol. 11, no. 26, 2024.

3 C. Jiang and H. Wang, “Network-aware value stacking of community battery via asynchronous distributed optimization.” in *Proceedings of the 2024 IEEE Power & Energy Society General Meeting (PESGM)*, Seattle, WA, USA, pp. 1–5, 2024.

4 E. Enasel and G. Dumitrascu, “Storage solutions for renewable energy: A review,” *Energy Nexus*, vol. 17, p. 100391, 2025.

5 A. B. Ahmad, “Renewable integration and energy storage management and conversion in grid systems: A comprehensive review,” *Energy Rep.*, vol. 13, pp. 2583–2602, 2025.

6 H. M. Ramos, “Energy storage systems in micro-grid of hybrid renewable energy solutions,” *Technologies*, vol. 13, no. (11), p. 527, 2025.

7 A. A. Alguhi and M. A. Alotaibi, “Optimal operation of battery energy storage systems in microgrid-connected distribution networks for economic efficiency and grid security,” *Energies*, vol. 18, no. 23, p. 6335, 2025.

8 R. Kandari, N. Neeraj, and A. Micallef, “Review on recent strategies for integrating energy storage systems in microgrids,” *Energies*, vol. 16, no. 1, p. 317, 2022.

9 A. N. Abdalla, “Integration of energy storage system and renewable energy sources based on artificial intelligence: An overview,” *J. Energy Storage*, vol. 40, p. 102811, 2021.

10 M. Y. Worku, “Recent advances in energy storage systems for renewable source grid integration: A comprehensive review,” *Sustainability*, vol. 14, no. 10, p. 5985, 2022.

11 M. Kiasari, M. Ghaffari, and H. H. Aly, “A comprehensive review of the current status of smart grid technologies for renewable energies integration and future trends: The role of machine learning and energy storage systems,” *Energies*, vol. 17, no. 16, p. 4128, 2024.

12 S. Sakib, M. B. Hossain, M. A. Zamee, M. J. Hossain, and M. A. Habib, “Role of battery energy storage systems: A comprehensive review on renewable energy zones integration in weak transmission networks,” *J. Energy Storage*, vol. 128, p. 117223, 2025.

13 R. I. Areola, A. A. Adebiyi, and K. Moloi, “Integrated energy storage systems for enhanced grid efficiency: A comprehensive review of technologies and applications,” *Energies*, vol. 18, no. 7, 2025.

14 G. M. Al-Amin, M. Shafiullah, M. Shoeb, S. M. Ferdous, and M. Anda, "Grid integration of EV: A review on stakeholders' objectives, challenges, and strategic implications," *e-Prime - Adv. Electr. Eng. Electron. Energy*, vol. 11, p. 100930, 2025.

15 X. Lei, J. Zhong, Y. Chen, Z. Shao, and L. Jian, "Grid integration of electric vehicles within electricity and carbon markets: A comprehensive overview," *eTransportation*, vol. 25, p. 100435, 2025.

16 S. Ali, et al., 2024. "Energy storage innovations in the context of electric vehicles and smart grid integration: A review," *Int. J. Econ. Env. Geol.*, vol. 15, no. 3.

17 National Renewable Energy Laboratory (NREL), "*Grid Congestion and Its Impact on U.S. Vehicle Electrification*," NREL Technical Report. Golden, CO: National Renewable Energy Laboratory, 2025.

18 International Energy Agency (IEA). "*Grid Integration of Electric Vehicles: A Manual for Policy Makers*," Paris: International Energy Agency, 2024.

19 kWh Summit, "The EV value stack: Beyond transportation," in *Proceedings of the 2025 kWh Summit*, 2025.

20 P. Kumar, "A comprehensive review of vehicle-to-grid integration in electric vehicles: Powering the future," *Energy Convers. Manage.:X*, vol. 25, p. 100864, 2025.

8

Conclusions and Future Directions

8.1 Summary of Key Concepts

The rapid advancement of transportation electrification and stationary energy storage has positioned electrochemical energy storage systems as a cornerstone of modern engineering. This book has systematically explored the fundamental principles, modeling techniques, charging strategies, and system level applications required to design safe, efficient, and sustainable energy storage systems for both mobility and grid-scale use.

8.1.1 Fundamentals of Lithium-ion Batteries

Lithium-ion batteries (LIBs) currently dominate the energy storage landscape due to their high energy density, high round-trip efficiency, and proven reliability. Their operation is based on the reversible intercalation and deintercalation of lithium ions between a graphite-based anode and a metal-oxide cathode, such as nickel manganese cobalt (NMC) or lithium iron phosphate (LFP). Battery performance is governed by internal electrochemical kinetics, including charge-transfer reactions and solid-state diffusion, which are strongly influenced by temperature, current rate (C-rate), and operating voltage limits.

Critical operational states, including state of charge (SOC), state of health (SOH), and state of power (SOP), cannot be measured directly and must instead be estimated using measurable electrical and thermal signals. Accurate estimation of these states is essential to ensure operation within the safe operating area (SOA), prevent degradation, and enable reliable energy and range prediction in real-world applications.

8.1.2 Alternative Storage and Hybrid Systems

While batteries offer superior energy density, alternative storage technologies provide complementary performance characteristics. Supercapacitors (or ultracapacitors) excel in power density and cycle life due to their electrostatic charge storage mechanisms in electric double-layer capacitors (EDLCs) and fast surface redox reactions in pseudocapacitors. Hybrid capacitor designs combine electrostatic and faradaic processes to bridge the gap between batteries and conventional capacitors.

Hybrid energy storage systems that integrate batteries with supercapacitors address the inherent trade-off between energy density and power capability. Such systems enable high peak power delivery, improved regenerative braking efficiency, and reduced stress on battery cells. Additionally, fuel cells, such as proton exchange membrane fuel cells (PEMFCs) and solid oxide fuel cells (SOFCs), offer efficient conversion of chemical energy (primarily hydrogen) into electricity, making them attractive for long-range, heavy-duty, and continuous-power applications.

8.1.3 Modeling and State Estimation

Effective battery management relies on accurate mathematical representations of battery behavior. Equivalent circuit models (ECMs), composed of resistive and capacitive elements, are widely adopted due to their balance between modeling accuracy and computational efficiency. These models form the foundation for advanced estimation techniques such as Kalman filters and their variants, which enable real-time estimation of SOC and other internal states under dynamic operating conditions.

Battery sizing and system design methodologies must simultaneously satisfy energy capacity and power capability requirements while accounting for thermal constraints, aging margins, and safety limits. These considerations are critical for ensuring reliable long-term operation in both electric vehicles and stationary energy storage systems.

8.1.4 Charging Systems and Infrastructure

Charging systems represent the critical interface between the electrical grid and energy storage devices. The constant-current constant-voltage (CC–CV) charging strategy remains the industry standard due to its simplicity and robustness. However, advanced charging techniques, such as multi-constant current (MCC) charging, pulse charging, and temperature-regulated charging have gained increasing attention for their ability to reduce charging time, mitigate thermal stress, and slow degradation.

Charging infrastructure spans residential Level 1 and Level 2 AC chargers to high-power Level 3 DC fast chargers. Interoperability and safety are governed by international standards such as CCS, CHAdeMO, and ISO 15118, which define electrical interfaces, communication protocols, and power negotiation mechanisms.

8.2 Challenges and Opportunities in Electric Energy Storage

Despite significant technological maturity, electric energy storage systems continue to face challenges that simultaneously create opportunities for innovation.

8.2.1 Degradation and Aging

Battery degradation remains a primary limitation, driven by mechanisms such as solid electrolyte interphase (SEI) growth, lithium plating, and mechanical stress from repeated cycling. Aggressive charging, high C-rates, and extreme operating temperatures accelerate these processes, leading to capacity fade and increased internal resistance. Addressing degradation presents opportunities for health-conscious control strategies, adaptive charging algorithms, and improved thermal management solutions that extend battery lifetime.

8.2.2 Thermal Management and Safety

Safety concerns, particularly thermal runaway, are inherent to high-energy-density lithium-ion systems. As charging power continues to increase, resistive heat generation (I^2R losses) places severe demands on cooling systems. Robust thermal interface design, advanced cooling architectures, and reliable state of temperature (SOT) estimation algorithms are essential to maintaining safe operation under fast-charging and high-power conditions.

8.2.3 State Estimation Complexity

Accurate estimation of SOC, SOH, and remaining useful life (RUL) is challenged by the nonlinear behavior of batteries and the flat open circuit voltage (OCV) characteristics of chemistries such as LFP. While model-based approaches provide strong physical interpretability, they rely on precise parameter identification. This challenge has driven growing interest in data-driven and hybrid estimation methods that leverage machine learning and cloud computing to enhance accuracy without excessive onboard computational requirements.

8.2.4 Grid Integration and Second-life Applications

The large-scale deployment of EV batteries raises significant end-of-life management challenges. However, retired EV batteries often retain 70–80% of their original capacity, creating substantial opportunities for second-life applications in stationary energy storage. Additionally, bidirectional charging and vehicle-to-grid (V2G) operation allow EVs to function as distributed energy resources, supporting grid stability, peak shaving, and renewable energy integration.

8.3 Future Directions in Energy Storage Research and Development

The future of energy storage is moving toward higher integration, intelligence, and sustainability.

1. **Advanced Materials and Chemistries:** Ongoing research focuses on silicon-rich anodes, cobalt-free cathodes, and alternative chemistries to improve energy density while reducing cost and reliance on critical raw materials. Solid-state batteries represent a major research frontier, offering the potential for enhanced safety, wider operating temperature ranges, and higher volumetric energy density through the replacement of liquid electrolytes.
2. **Smart and Cloud-connected Battery Management Systems:** Next-generation Battery Management Systems are evolving from standalone embedded controllers to intelligent, cloud-connected platforms. AI-driven BMS architectures enable the creation of digital twins for individual batteries and for entire fleets, facilitating predictive maintenance, long-term health forecasting, and real-time operational optimization based on large-scale data analytics.
3. **Hybrid and Multi-physics Energy Systems:** Future energy storage and propulsion systems will feature deeper integration of multiple technologies. Hybrid battery–supercapacitor systems will be increasingly adopted to optimize power delivery and efficiency, while compact multi-functional architectures, such as integrated e-axles combining motors, inverters, and charging electronics, will continue to improve system-level performance and power density.
4. **Circular Economy and Sustainability:** As the volume of retired batteries grows, circular economy considerations will become central to energy storage development. Standardized testing, grading, disassembly, and repurposing techniques will be essential for second-life deployment, while advances

in recycling technologies will help recover valuable materials and reduce environmental impact. Closing the battery lifecycle loop will be critical for long-term sustainability.

8.4 Concluding Remarks

Electric energy storage has evolved from a supporting technology into a foundational pillar of modern energy and transportation systems. The concepts, methods, and applications presented throughout this book demonstrate that effective energy storage design requires a multidisciplinary approach, integrating electrochemistry, electrical engineering, control systems, and power systems engineering. As electrification and renewable energy penetration continue to expand, the importance of robust, efficient, and sustainable energy storage solutions will only grow. The challenges ahead are significant, but they are matched by equally significant opportunities for innovation and global impact.

Index

Note: Page numbers in *italics* and **bold** refers to figures and tables respectively.

Wiley The manufacturer's authorized representative according to the EU General Product Safety Regulation is Wiley-VCH GmbH, Boschstr. 12, 69469 Weinheim, Germany, e-mail: Product_Safety@wiley.com.

Printed and bound by CPI Group (UK) Ltd, Croydon, CR0 4YY
02/10/2026
02227444-0002